AF760798

Virtual Reality

Virtual Reality

Representations in Contemporary Media

Melanie Chan

B L O O M S B U R Y
NEW YORK • LONDON • NEW DELHI • SYDNEY

Bloomsbury Academic
An imprint of Bloomsbury Publishing Inc

1385 Broadway
New York
NY 10018
USA

50 Bedford Square
London
WC1B 3DP
UK

www.bloomsbury.com

Bloomsbury is a registered trade mark of Bloomsbury Publishing Plc

First published 2014

© Melanie Chan, 2014

All rights reserved. No part of this publication may be reproduced or transmitted in any form or by any means, electronic or mechanical, including photocopying, recording, or any information storage or retrieval system, without prior permission in writing from the publishers.

No responsibility for loss caused to any individual or organization acting on or refraining from action as a result of the material in this publication can be accepted by Bloomsbury or the author.

Library of Congress Cataloging-in-Publication Data
Chan, Melanie.
Virtual reality : representations in contemporary media / Melanie Chan.
pages cm
Includes bibliographical references and index.
Includes filmography.
ISBN 978-1-4411-7531-1 (hardback : alk. paper) 1. Virtual reality in mass media.
2. Mass media and technology. 3. Virtual reality–Social aspects. I. Title.
P96.V55C53 2014
006.8–dc23
2013035151

ISBN: HB: 978-1-4411-7531-1
EPub: 978-1-6235-6474-2
EPDF: 978-1-6289-2147-2

Typeset by Integra Software Services Pvt. Ltd.

Contents

Acknowledgements

I would like to express deep gratitude for the support received from colleagues at Leeds Metropolitan University. Indeed, the enthusiastic support and collegial atmosphere within the university considerably enhanced the writing of this book. In addition, I am extremely grateful to my doctoral studies supervisors, Graham May, Prof. Guy Julier and Stephen Rennie, for their valuable insights and encouragement. Some of the ideas expressed in the book were tested out and expanded through papers presented at various conferences such as the International Communications Association conference in Montreal in 2008. I give special thanks to the British Academy for their generous financial support, which enabled me to attend this conference and discuss my ideas with other scholars in the field of media, communication and cultural studies.

I also want to express my thanks, appreciation and indebtedness to film scholar María Seijo Richart for her support and encouragement over the many months that it took to produce this book. Dr Simon Blanchard and Dr Gil Dekel also provided invaluable assistance and feedback on the manuscript. Finally, thanks also go to the editorial team at Bloomsbury, especially Katie Gallof, for their support through the publishing process. I would also like to express my enduring gratitude, respect and love to Michael Chan, who provided support, encouragement and editorial support over many months and years. Indeed, his input helped to catalyse the ideas presented in the book.

Introduction

This book places the study of representations of virtual reality in contemporary media within a historical, social and cultural context. In doing so it indicates that virtual reality has a long and complex historical trajectory, from the use of hallucinogenic plants to visual styles such as Trompe L'Oeil and more recent developments such as computer games. As this book indicates, many different layers of meaning have become encrusted around the term 'virtual reality'. Firstly, the term can be used to refer to computer-generated environments that involve some form of interactivity and immersion. In this context interactivity is a form of reciprocal engagement with a virtual environment, whilst the term 'immersion' refers to the feeling of being encompassed by computer-generated imagery (Sohn, 2011). So instead of viewing an image with a defined boundary such as a frame, virtual reality seems to offer the possibility of moving beyond the frame and into the image. Secondly, the juxtaposition of the terms 'virtual' and 'reality' can be used as a starting point for the consideration of philosophical debates regarding contemporary media representations and simulation. Therefore, virtual reality does not just refer to computer-generated spaces; rather, taken more broadly, it can be regarded as something that is almost but not quite real or as a potential that has yet to become actualized (Deleuze, 1997). Taking these two approaches as a starting point, this book aims to explore virtual reality both in relation to technology and also in relation to debates about what constitutes the real and the virtual and how these concepts are played out in contemporary media representations.

To limit the scope of the discussions arising from the virtual and the real, this book pays particular attention to media representations of virtual reality in the 1980s and 1990s – for what is striking about this period is that virtual reality was often represented as a wondrous technology that could provide an opportunity to transcend the limitations of physical embodiment. Therefore, the purpose of this book is to critically analyse contemporary media representations of virtual reality, paying particular attention to the ways in which they deny, repress or overlook the embodied dimension of our experiences. In doing so, I will contend that embodiment is a fundamental aspect of immersion in virtual reality rather than something to be transcended. In this way, the book aims to challenge misleading ideas about transcendence and productively contribute to debates about the relationships between the virtual and the real.

In this book, I insist that contemporary media representations of virtual reality are incredibly influential since they can shape our attitudes towards the virtual and the real. Consequently, this book aims to raise a series of questions about the ways in which representations are produced, how they are circulated and the meanings they engender – for, as this book indicates, contemporary media representations of virtual reality generate clusters of meaning in relation to perception, embodiment and immersion. These meanings are circulated, perpetuated and challenged in a wider social and cultural context. This book also contributes to this process of negotiation and debate. However, it does not attempt to pin down meaning; rather it embraces multiplicity and openness regarding what constitutes the virtual and the real. Through taking this approach, I propose that embodiment is an interactive and dynamic process that involves the spaces, virtual or otherwise, with which we engage.

In the 1980s computer technologies were not yet a major aspect of our daily lives. Back then computers were huge machines that were used for the purposes of university research or were to be found in the offices of major corporations. However, the computer games industry was starting to become established and arcade machines introduced a new generation to computer-generated virtual spaces. It was during this period that William Gibson wrote his influential novel *Neuromancer* (1984), which represented cyberspace as a new computer-generated space where conflicts, battles and adventures took place. It was also in the early 1980s that computer scientist Jaron Lanier was credited with coining the term 'virtual reality'. To begin with, Lanier was enamoured with the term because it was quirky and contradictory. However, in an interview with Oliver Burkeman in 2001, Lanier claimed that virtual reality has become problematic due to the hype and overly optimistic expectations that are associated with this term. Indeed, as scholar and journalist Erik Davis points out, the term 'virtual reality' acquired cultural currency during the 1980s and 1990s: 'hitting the mass brainstem like a rush of crack, the term rapidly took on the millennialist charge of all pop futurisms' (1999, 190).

By the 1990s virtual reality was particularly nascent in relation to economic and political debates about information technology (Gore 1991) since personal computers and the Internet became commercially available, providing access to computer-generated virtual worlds and to communication mechanisms such as email. It was also during this period that popular film and literature represented strange, exotic and dangerous immersive virtual worlds.

By the 1990s hermeneutic studies and popular non-fiction publications also explored various aspects of virtual reality technology. In 1991, *Virtual Reality* was published by the cultural critic Howard Rheingold and this provided a historical survey of the development of this technology, mostly in relation to university research groups, technology-based companies and computer specialists in California (particularly in the region known as Silicon Valley). Two years later, media journalist Benjamin Woolley provided a thoughtful and provoking discussion of the cultural, social and philosophical ideas surrounding virtual reality in *Virtual Worlds* (1993). Another significant publication during this period was Michael Heim's *The Metaphysics of Virtual Reality* (1994), which explored the philosophical aspects of the virtual and the real, whilst *Cultures of the Internet: Virtual Spaces, Real Histories, Living Bodies* (1996), edited by Rob Shields, focused on such topics as multi-user dungeons, free speech and the Internet. A further significant exploration of virtual reality during the 1990s was *Digital Sensations* (1999) by Ken Hillis. Indeed, *Digital Sensations* provided valuable critical analysis of the relationships between embodiment and immersion in virtual reality. Finally, *Virtual Geographies: Bodies, Space and Relations* (Crang, M. Crang J. and May, 1999) provided valuable insight into the relationships between embodiment and virtual reality. Since the millennium, critical commentary on virtual reality has continued. For instance, *Digital and Other Virtualities: Renegotiating the Image* (2010), edited by Griselda Pollock and Antony Bryant, presents a thorough overview of the symbolic aspects of the virtual within visual culture. Meanwhile media artist and theorist Frances Dyson (2009) provides a thorough and considered account of sound and virtual reality in *Sounding New Media: Immersion and Embodiment in the Arts and Culture*.

Although debates about virtual reality were particularly prominent during the 1980s and 1990s, it has been argued that there is now diminished interest in this technology within mainstream popular culture. For instance, Joanna Buick claims that 'the general public, having been bombarded with a frenzy of hype about VR and shoot-em-up computer simulation games in the last couple of years, are already disappointed – they think it's all over' (2002, 110). On the one hand, Buick was right to point to the initial hype surrounding virtual reality and that initially computer-generated 'shoot-em-up' games found in malls such as Trocadero in London were of poor quality, in terms of refresh rates and graphic resolution, and this could have led to some disillusionment with this technology. On the other hand, Buick's remarks were made over a decade ago and it could be argued that virtual reality is now more commonplace because it is embedded within industrial or military applications such as the aerospace and automobile industry. For example, SGI Japan produces high-specification virtual reality systems that are used for clients such as Komatsu, for use in the construction industry, whilst Sim Industries (a Lockheed Martin Company) produces virtual reality systems for the aerospace industry. So, whilst virtual reality technology may not appear to be as prominent as it was during the 1980s and 1990s it has become mainstreamed in industrial and military applications and less of a technology novelty.

Now, in the second decade of the twenty-first century, at a time when computerized digital technologies permeate our everyday lives, it is timely to investigate the concepts of the virtual and the real and how they gained traction. In doing so, I will take heed of the warning proclamations of French theorist Jean Baudrillard, who wrote:

> Enquiry into the virtual is made even more delicate and complex today by the extraordinary hype surrounding it. The excess of information, the massive advertising effort, the technological pressure, the media, the infatuation or panic – everything is contributing to a kind of collective hallucination of the virtual and its effects. (2002, 111)

At present we can simultaneously navigate the material spaces around us as well as the virtual online spaces through mobile technologies such as smart phones or tablet computers. Taking this into consideration, William Gibson's assertion that we are now living in cyberspace seems appropriate (Gibson in Wallace-Wells, 2011). Therefore, this book aims to provide a major and timely retrospective of the representation of virtual reality as a technology and as a series of philosophical and cultural ideas within contemporary media. In particular, it analyses immersion and interactivity in relation to recent developments in 3D film as well as multi-user online environments such as *Second Life* and *World of Warcraft*. By taking an interdisciplinary approach, the book attempts to offer a comprehensive and critical account of the relationships between embodiment and technology.

This book follows cultural theorist Mark Hansen's (2004) argument that we do not just see virtual environments; we experience them through the body.

Furthermore, Heather Brooke (2011) recounts that the word 'digital' stems from the Latin term *Digitus*, which refers to fingers or toes. As people used their fingers or toes to count (ten fingers were referred to as digita) the digital became associated with numerals and mathematics. Therefore, embodiment plays an important part in our understanding of digital technologies such as computer-generated virtual environments. But crucially our engagement with technology is altering our sense of embodiment and how we experience the world around us. Indeed, the work of cultural geographer Nigel Thrift (2004) indicates that our sense of touch is changing due to our interaction with technologies, such as using keyboards, keypads, the mouse and the remote control. These changes can be detected in relation to systems such as the Nintendo Wii™, in which our embodied interaction is shaped in relation to the use of a control pad and the images we interact with on screen.

In addition, this book explores the embodied material conditions in which virtual reality is produced, circulated and consumed using a range of theoretical models, from phenomenology to contemporary media studies. Setting out the critical terrain, Stephen Talbott (1995) claims that the problem with virtual reality is that it has become entangled with the idea that computers can improve our lives. In this way, social issues are transformed into technological ones. In fact, in some cases biological metaphors based on reductive interpretations of Darwin's theory of evolution are used as a means of referring to technology. Outlining this process, Talbott writes, 'social evolution is patterned after technological development, wherein newer, better, more sophisticated products inevitably replace older, more primitive ones' (1995, 5). Joan Broadhurst Dixon and Eric Cassidy provide another viewpoint on these issues through their observation that debates about new technologies can often become reductive and polarized. Consequently 'those who are against technologies are regarded as having "paranoid technophobia" while the pro-techies are aware of the coming information revolution' (Broadhurst Dixon and Cassidy, 1998, preface).

In a way, there is nothing particularly unique about arguments concerning the transformational possibilities of virtual reality technology. For instance, the development of the steam train and the automobile also radically altered our sense of embodiment in relation to space, time and speed. In addition, arguments concerning disembodiment have also been made in relation to the radio, which transmitted the voice over vast distances, or the idea that the tape recorder severed speech from the living presence of the speaker (Hayles, 1999). Moreover, as Thrift points out, over time our environment has become more artificial:

> Roads, lighting, pipes, paper, screws and similar constituted the first wave of artificiality. Now a second wave or second artificiality is appearing, extending its fugitive presence through object frames as diverse as cables, formulae, wireless signals, screens, software, artificial fibres and so on. (2004, 585)

Arguably, the second wave of artificiality identified by Thrift has become so commonplace that we barely acknowledge it. For navigating digital terrains through the use of the World Wide Web, social networking and smart phone technology have become part of our everyday lives. But rather than taking technology for granted because it is so ubiquitous, it is important to critique the specific conditions in which different technologies are produced and utilized. Indeed, embodiment is a key issue in relation to the social and material conditions surrounding the production and consumption of technology. Therefore, the debates surrounding virtual reality go beyond technology towards a wider set of concerns relating to shifting relationships concerning embodiment, immersion and engagement with the world around us. This does not mean that such technologies simply determine our conception of reality or fundamentally change our attitudes towards embodiment. Instead, as this book indicates, what constitutes the virtual, real and embodied experiences are contested terrains that shift and change within different social, cultural and historical contexts.

In Chapter 1 the historical strands surrounding the development of virtual reality are briefly outlined in order to set the context for later chapters in the book. The chapter also sets the terms for a critical analysis of the debates surrounding the virtual and the real. Specifically in this chapter I insist that the idea that virtual reality is a radically new emergent technology is somewhat misleading. Instead, virtual reality is presented as part of a long but not unbroken historical trajectory. Following this line of thinking, the chapter provides a critical overview of the ways in which virtual reality draws upon developments within the field of computing, from the mechanical computational devices that were used in the nineteenth century to the present-day era of the World Wide Web, computer games and 3D film.

In Chapter 2, I interpret virtual reality through the lens of simulation theory, particularly the work of Jean Baudrillard, Sean Cubitt and Sadie Plant. I then take some of the key ideas that are associated with simulation and apply them to specific examples. Drawing upon the work of Doel and Clarke (1999) I also explore virtual reality through a four-fold schema. Firstly, I explore the idea that the virtual is a degraded copy of the real. Secondly, I will consider how the virtual can augment or supplement embodied experience. Thirdly, I also need to think carefully about the idea that in contemporary culture the virtual is becoming indistinguishable from reality. Finally, I will explore the possibility that virtual reality generates its own reality, thereby becoming hyperreal.

Chapter 3 considers the ways in which popular film can open up important critical debates about technology and embodiment. In the 1980s and 1990s there was a plethora of popular films that present us with ambivalent representations of spectatorship, embodiment and mediated experiences, including *Tron* (1982), *Total Recall* (1990), *The Lawnmower Man* (1992) and *Strange Days* (1995). But this chapter focuses particularly on *Strange Days* since this film concerns the use of a fictional device which enables one person's experiences to be recorded and played back by someone else. These playback recordings are bought and sold as commodities which provide intense and compelling immersive experiences. In addition, the ways in which *Strange Days* utilizes point-of-view (POV) camerawork raises a series of important questions about mediated and immersive experiences by inviting spectators to form an intimate and intense connection with the characters in the film and their experiences.

Chapter 3 discusses the ways in which *Strange Days* is set in Los Angeles at the turn of the millennium at a time of civil unrest and social estrangement. These social tensions contribute to the appeal of privatized mediated experiences in order to escape the potential dangers of social interaction. When *Strange Days* was released in 1995 it offered a disturbing vision of the future, at the turn of the millennium. Yet in Chapter 3 we will see that the ways in which the film represents playback technology as a way of accessing compelling immersive experiences remain relevant today. Indeed, these are interesting points to consider at a time when technologically mediated communicative experiences are now becoming commonplace through the widespread use of the World Wide Web. Moreover, technologically immersive experiences have become pervasive and commercially accessible through the use of gaming consoles such as Microsoft X-Box, Nintendo Wii™ and the Sony Playstation®. Another more recent development is augmented reality, whereby digital imagery can overlay our experience of the world around us. For instance, Google has produced a device that records what the wearer sees and their visual experiences can be shared instantly through links to social networking technology, a development that echoes the representation of playback recordings in *Strange Days*. Therefore, by undertaking an analysis of *Strange Days*, Chapter 3 aims to provide a springboard for the discussion of contemporary debates concerning technologically mediated and immersive experiences.

Building on some of the groundwork concerning technologically mediated experiences in the previous chapters, Chapter 4 explores the relationships between online and offline identities. Indeed, this chapter raises questions about whether it is still relevant to make clear distinctions between them. In the mid-1990s computer-generated worlds were accessed through desktop computers and modems which were fixed at a particular location and therefore it made sense to refer to online and offline identities because the two realms seemed distinctive. But now we can explore the digital realms of the World Wide Web or computer games whilst we move through physical space through such devices as tablet computers or smart phones. As Nick Harkaway (2012) observes, mobile devices have become portals that enable us to become immersed and engrossed within digital environments. So at a time when we can simultaneously traverse digital realms and the world around us, it is opportune to reflect upon the distinction between online and offline identities. To explicate these ideas further there will be an in-depth analysis of three contemporary films – *Catfish*, *Surrogates* and *Gamer* – which point to some of the tensions that arise from navigating online and offline identities.

Chapter 5 scrutinizes the ways in which immersion in virtual reality is associated with technological transcendence. In this chapter I will contend that the hype and hope that are associated with transcending the physical body for exalted wonderment in virtual realities can be traced back to the seventeenth century, to the dualistic philosophy of René Descartes. I also make evident that visions of transcending the limitations of physical embodiment can be detected in such films as *Brain Storm*, *Ghost in the Shell* and *Source Code*. On the surface these films can be interpreted as presenting an escapist fantasy whereby the limitations of physical embodiment can be transcended through immersion in virtual reality technology. But this chapter will assert that such ideas are problematic, especially when they are based on a misleading division between the mind, brain and physical body. To investigate the ways in which technological development intersects our understanding of the relationships between embodiment and immersion in virtual reality, the chapter includes references to information theory and cybernetics. In particular, it explores the ways in which Claude Shannon's framework of logical propositions and signal processing has been applied to human cognition. For instance, it considers what happens when cognitive processing is regarded as a pattern of signals that can be mapped and recreated, and how this relates to visions of transcending the human body for a technological afterlife in the work of computer scientist

Ray Kurzweil. To illuminate these issues further the chapter considers the relationships between dualistic approaches to the human body and visions of a technological afterlife in the film *Source Code*.

Chapter 6 explores the relationships between technology and the phenomenological concept of being-in-the-world. This is achieved through a discussion of *Osmose*, which is an installation art piece produced by Char Davies and a team of technological specialists, and the film *Avatar* (2009, dir. James Cameron). This chapter indicates that *Osmose* was produced in the mid-1990s at a time when representations and debates about virtual reality were foregrounded. Consequently, *Osmose* generated interest in the creative possibilities of virtual reality as a new artistic medium. What will also be shown to be striking about *Osmose* is that navigation through this virtual world involves an intimate connection with the body. Indeed those who became immersed in *Osmose* were required to wear a cyber-vest that monitored the movements of the body, which formed a feedback loop between them and the virtual world.

When *Avatar* was released the film received a great deal of popular and critical attention due to the ways in which it utilized 3D technology to provide an immersive and captivating cinematic experience. However, the chapter will consider the ways in which *Avatar* can be regarded as utilizing 3D technology to stimulate the film industry at a time when it was threatened by other media forms such as computer games. Although there are differences in terms of the production and consumption processes surrounding *Osmose* and *Avatar*, this chapter indicates that they draw upon similar symbolic motifs in order to provoke important questions and concerns about technology. In this regard, the chapter points to the argument that it seems churlish to criticize the technological wonders that are found in contemporary society, especially through highly technological forms of cultural production such as *Osmose* or *Avatar*.

The final chapter (Chapter 7) considers the ways in which the novel *Zero History* (2010) by William Gibson opens up a series of concerns about the relationships between symbolic systems of signification, the virtual and the real. It focuses on passages in *Zero History* that refer to the concept of the simulacrum, which is a sign that gains meaning through reference to another sign within a self-referential signifying system (Baudrillard, 1983, 1998). In doing so, we will consider how the simulacrum instigates a situation in which meaning skims across the surface of different signs occluding the historical and material context in which they are produced, circulated and understood. The chapter also pinpoints the ways in which *Zero History* represents fashion and design as signifying codes that circulate in a rapid self-referential global system that is replete with brand names. Extending the debates surrounding signification and self-referentiality, this chapter also considers the intertextual dimensions to Gibson's oeuvre. Drawing upon postmodern theory, Chapter 7 considers whether we are now living in a culture of virtuality because signs have become increasingly ludic, de-stabilized and free-floating. But in order to consider this argument from another perspective, this chapter also points out that the idea of free-floating signs has actually arisen from specific social and economic conditions (Jameson, 1984, Dunst 2008).

Chapter 7 indicates that *Zero History* draws our attention to the complex, multifaceted levels of interpretation within literary expression and how this relates to changing relationships between signs and meanings. *Zero History* also allows us to explore some of the differences between literary expression and computer code – for computer code is a signifying system that strives to achieve a fundamental correspondence between signs and meaning and ambiguity is something to be eradicated since this can lead to system errors (Hillis, 1998). Conversely, the creative realms of fashion or literature thrive on ambiguity, on the twisting and turning of structures of meaning. So in order to critically analyse the representation of signs and meanings in *Zero History* in more detail, I will draw upon the work of Katherine N. Hayles (2005), paying particular attention to the distinction she makes between the operational parameters of computer code and literary expression.

It also seems appropriate to end the book with a discussion of *Zero History* since Gibson is a writer who is associated with representing cyberculture and virtual realities from the 1980s onwards. But in recent years Gibson has moved away from futuristic science fiction visions of technologically saturated culture. Now Gibson's work provides a subtle yet equally powerful representation of the shifts that have occurred in the realm of symbolic signification and the relationships between the virtual and the real. Furthermore, as this book indicates, the initial exuberance and wonderment that was associated with immersive virtual reality in the 1980s and 1990s seem to have faded. Moreover, the discursive arena seems to have shifted away from cyberspace, cyberpunk and cyberculture. Instead, other aspects of digital culture have entered the discursive arena, such as web 2.0, social networking and Big Data. Yet it is precisely because navigating computer-generated spaces and communicating with one another via digital devices have become commonplace that it is necessary to step back and assess how this relates to conceptions of the virtual and the real and how such ideas are represented in contemporary media.

1

Histories of Virtual Reality

This chapter has two broad aims: firstly, it explores various historical strands surrounding the development of virtual reality, and secondly it critically analyses debates surrounding the virtual and the real. The purpose of this two-fold approach is to contextualize the discussion of contemporary media representations of virtual reality in subsequent chapters. In doing so, I will place the discussion of contemporary representations of virtual reality within a wider historical, social and cultural context. However, there are difficulties with romantic notions of constructing the history of virtual reality through the accumulation of archive materials (Steedman, 2005), for the immense amount of information and interrelations that take place even during a specific historical moment far exceeds the archival process. Therefore, at the outset it is admitted that this chapter presents a highly truncated version of the history surrounding virtual reality. But my main purpose in this chapter is to challenge the idea that virtual reality is a radically new technology by indicating that it has a long, complex and uneven trajectory.

To make the discussion manageable it focuses on two key thematic strands: computation and cyberculture. The first strand, computation, explores the ways in which virtual reality environments are made possible by computerized algorithmic processing. In particular, there will be an exploration of the development of computing as a form of mechanized calculation and how this relates to changes in social production and the displacement of human labour. In this regard, the discussion will illustrate the historical context in which machines became associated with a particular form of intelligence and the ideologies that underpinned this viewpoint. Commenting on this point, Simon Schaffer states that 'to make machines look intelligent it was necessary that the sources of their power, the labour force which surrounded and ran them, be rendered invisible'

(1994, 204). Following on from this point regarding invisibility, this chapter will indicate that whilst electronic devices such as computers or the software that generate virtual reality environments are often represented in contemporary media as desirable consumables, the ways in which they are produced is largely hidden from view. So the human labour that surrounds the production of technological devices in factories in South East Asia is occluded in favour of alluring advertisements that promote technological devices and services. In this way, the historical discussion of the development of mechanized calculation provides the foundation for the subsequent exploration of the relationships between technology and embodiment and also for the understanding of virtual reality in the present day.

In the second strand of the discussion attention will be given to the ways in which developments in photography, film and computer-generated imagery relate to the emergence of cyberculture and contemporary forms of virtual reality. Specifically, there will be an outline of the work of cinematographer Fred Waller, who produced a three-dimensional cinematic experience, known as Cinerama, and Morton Heilig's multi-sensory simulator, Sensorama. These cinematic explorations will then provide the backdrop for the understanding of contemporary media representations of virtual reality. In addition, there will be an exploration of computer games since these are now a significant part of the entertainment industry, with revenues that often eclipse the box office revenue of major Hollywood films. Although there is not a long, historical body of academic scholarship on computer games, in recent years dedicated journals on gaming and simulation have been established. At the same time, it is becoming increasingly difficult to precisely define what constitutes a computer game, since there are some simulations and game applications that are open-ended and do not operate according to a pre-defined established set of rules, such as *Second Life* or *The Sims* (Corliss, 2011).

In recent years contemporary computer games such as *The Sims 3*, *FIFA 10* and *Battlefield: Bad Company 2* (Electronic Arts) and *Final Fantasy*® and *Tomb Raider*® (Eidos) have become incredibly popular, often selling millions of units. Additionally, force feedback devices have now become commercially available and increasingly popular. This includes devices such as the Nintendo Wii, which was launched in October 2006, the SideWinder by Microsoft and the Driving Force™ steering wheel by Logitech, which can be used with Sony Playstation games such as *Gran Tourismo*. With recent developments such as the Kinect, which is used with Microsoft's X-Box 360, users can now interact with a suite of multimedia forms (games, music and film) through gestures and voice recognition. Therefore, the user's body becomes the control device, the tool that allows interaction between them and the X-Box device. From these examples, we can see that human–computer interaction with virtual worlds has become popularized through computer games and is now a lucrative commercial area within a global capitalist economy. But to begin with, we will take a step back from contemporary digital technologies and briefly explore some of the historical, social and cultural factors surrounding their development.

Computation

Historically, connections can be drawn between virtual reality technologies, textile production and mechanical forms of computation (or calculation). Joseph Marie Jacquard, a French engineer, first developed an automated machine for weaving thread. Other developments in the silk industry such as the punch card pattern system designed by Basil Bouchon circa 1725 contributed to later developments such as the card machine which was used to tabulate US Census Data. For instance, the card machine developed by Herman Hollerith was first used to process the 1890 Census and this reduced the time taken to process the data from a matter of years to just a few months. By the nineteenth century the introduction of the mechanized loom also had a major impact on the production of textiles. Specifically, small-scale textile production was displaced by large-scale industrial modes of production in mills and factories. In addition, these changes altered the ways in which human skill and intelligence were measured and evaluated.

During the nineteenth century, Charles Babbage (1791–1871) and Ada Lovelace (1815–1852) developed a mechanized system for mathematical processing, which altered social modes of production. However, Herbert Klein points out that 'most accounts of the development of modern computing deny a straight connection between Babbage's mechanical device and electronic computers. It is therefore not so much a direct line of descent but rather a common heritage' (2007, 37). Of note is that in Babbage's time routine calculations were a form of clerical work that was performed by human labourers known as computers (Schaffer, 1994). So computation is not something that simply concerns digital technology or contemporary forms of computing machines (desktop machines, laptops and tablets) since other methods can be used to perform such calculations (Hayles, 2005).

Babbage aimed to produce a machine that was quicker and more efficient at performing calculations than the mental labour of human beings. According to Klein, 'Babbage believed that the human mind worked on the same mechanical principles as a machine, so that it could also be emulated by a machine' (40). Babbage produced the *Difference Engine* (1820–1822), which performed mathematical operations and could be used for performing the calculations that were required for navigational and astronomical tables. Later he worked on a device called the *Analytical Engine* (1833–1871), which utilized a series of punched cards, like an early form of computer programming. Klein states that this was 'an idea which Babbage had taken from Joseph Jacquard, inventor of the automatic loom' (39). In a similar way to contemporary electronic computers, the *Analytical Engine* aimed to be a universal machine which could be programmed to execute many different functions. Although Babbage spent many years working on the *Analytical Engine*, he was not able to produce a fully functioning machine during his lifetime.

At the same time that Babbage was working on replacing mental labour using the *Difference Engine*, many other parts of the country, especially Manchester and Leeds, were the site of industrial production and gruelling human labour in mills and factories. The operation of machines could be measured and monitored and soon these practices were applied to the human labour that took place in mills and factories. Workers were regarded as machine-like and their endeavours were carefully programmed into a production system that comprised of different components or stages. Schaffer contends that 'under Babbage's productive gaze, the powers of the body were rendered mechanical and thus profitable or wasteful and thus consigned to oblivion' (1994, 227). So Babbage's vision of replacing mental labour was soon applied to the physical aspects of material production as well, through the mechanization of work and systems of measurement and surveillance. Klein states that during his lifetime few people recognized Babbage's vision and his accomplishments and that he was the subject of ridicule. Yet Babbage's vision of a universal machine and the replacement of human labour remain important today. It could be argued that this process has continued into the twenty-first century as the labour force surrounding the production of the computerized components that drive virtual environments in factories in South East Asia is largely hidden from view, a point that will be discussed in more detail as this chapter unfolds.

As discussed, nineteenth-century developments in mechanized calculation paved the way for the production of electronic computers in the twentieth century. However, further developments were also required, such as the electrical transistor, which enabled algorithmic calculations to be performed at greater speeds. With the development of the Soviet Space programme after the end of the Second World War, the American military began to invest heavily into research and development, particularly in the field of computing. Within the context of Cold War relations in the 1950s and 1960s there was a sense of urgency within the US military regarding technological development, which led to experimentation with interactive computing and simulation, often via the funding of university research projects. Moreover, as Howard Rheingold (1991), Benjamin Woolley (1993) and Ken Hillis (1999) have pointed out, the military facility known as the Advanced Research Projects Agency (ARPA) was a major funding source for academic research into virtual reality. Douglas Engelbart, a former radar technician in the Second World War, received ARPA funding to establish the Augmentation Research Center (ARC) at Menlo Park, California, for Stanford University. The remit of ARC was to focus on the development of computer visualization systems. What is significant for the purposes of this study into virtual reality is that the research conducted at Menlo Park indicated the possibilities of human and computer interaction. Indeed, Engelbart envisaged that computers could be used to extend human capabilities and worked on the development of information displays on computer screens, which was a somewhat radical idea during this period (Burnett, 2004).

It was also during the 1960s that a research student named Ivan Sutherland at the Massachusetts Institute of Technology (MIT) produced a rudimentary form of computer graphics called Sketchpad and a head-mounted display device. Sutherland's prototype graphics system was innovative, exciting and groundbreaking because it indicated the creative potential of computer graphics. Explaining how Sketchpad operated, Howard Rheingold tells us that 'people could create images in the most natural way possible, by using their hands and eyes and a pen-like device to draw them' (1991, 90). Sutherland's early experiments into computer graphics are significant because they provided the foundation for the graphical interfaces that are used for contemporary forms of virtual reality technology. In 1965 Sutherland presented a research paper entitled 'The Ultimate Display', which provided an account of his experimental work on head-mounted displays and virtual reality systems. In this paper, Sutherland discusses the ways in which we perceive and interpret the world around us via the sensory motor capabilities of the human body and our prior knowledge of objects. Of note is that Sutherland's research focused upon the embodied aspects of immersion in virtual reality. His research suggests that in order for a computer-generated environment to be effective it needs to engage as many senses as possible, rather than concentrating solely on sight. According to Sutherland, his Ultimate Display would employ the use of joysticks, stereophonics and force-feedback mechanisms to maximize the immersive experience. The drawback was that the prototype head-mounted display that Sutherland developed was extremely cumbersome and uncomfortable to wear. By the late 1960s Sutherland moved to the University of Utah and after further refinements to his research developed the first operational head-mounted display system.

A further strand to the historical development of virtual reality relates to experimentation with hallucinogenic drugs and alternative states of consciousness in the 1960s. Human beings have attempted to explore other realities throughout the ages, through creative forms, the use of hallucinogenic plants or rituals involving music and dance. In some cases, only certain members of society were considered as worthy of exploring other realities and aspects of consciousness such as shamans or mystics. In the 1960s, however, it was thought that LSD could provide a chemical basis for what were considered mystical or spiritual experiences. In the United States research into LSD was conducted in academic circles, at Harvard and Berkeley, and there is a connection between these experiments and the development of virtual reality environments. Timothy Leary, a major proponent of virtual reality, experimented with hallucinogenic drugs as a means of accessing alternative realities when he was a professor at Harvard.[1] In the following statement, Leary outlines the connections between research into LSD and the development of personal computing:

> In 1976, the Apple computer was introduced. At the same time video games provided young people with a hands-on experience of moving flashy electronic, digital information around on screens. It was no accident that many of these early designers and marketers of these electronic appliances lived in the San Francisco area and tended to be intelligent adepts in the use of psychedelic drugs. (1994, 40)

Leary makes the point that screen technologies such as television, video and computing have to some degree altered what is meant by the term 'reality'. When interpreted in this context, virtual reality is not about passively inducing alternate states of consciousness through drugs but about interactive engagement with images and sound.

[1] The research Leary conducted into LSD led to his removal from his Professorship post at Harvard. His colourful life is outlined in his obituary in the *New York Times* of 1 June 1996. See http://www.nytimes.com/learning/general/onthisday/bday/1022.html

By the late 1980s and throughout the 1990s representations of virtual reality technology had a particularly high profile, in Western and Westernized culture. In the political arena, Vice-President Al Gore championed the development of the Internet, World Wide Web and virtual reality, stating that 'we need to make sure that the federal government does all it can to stimulate innovative and truly important new technologies like virtual reality' (Gore, 1991). Of particular note is Gore's promotion of the concept of the information superhighway and the development of a knowledge-based economy which was also linked to the virtual reality technology. In contrast to Gore, Timothy Leary was more at the fringes of the vectors of social power and prestige, promoting virtual reality through association with writers such as William Gibson, William S. Burroughs and John Perry Barlow (formerly of the group 'The Grateful Dead'). According to Leary (1994), the development of computer-generated virtual reality marked a breakthrough which would significantly shift our understanding of embodied relationships, time and space. As discussed, the statements made by Gore and Leary provide a useful starting point to open up further debates about the construction and circulation of representations of virtual reality and how they operate within networks of socio-economic power and within a cultural context.

Cybernetic totalism

In *The Third Culture*, Jaron Lanier (2000) uses the term 'cybernetic totalists' to describe those intellectuals, scientists, researchers and developers who promote the notion of a digital utopia which can be achieved through computing technology.[2] Lanier does not deny that computing and information technology have changed the ways in which we think and communicate; however, he is cautious about claims of universality with respect to computer development. Lanier states that cybernetic totalists claim that computing is a global phenomenon, yet some countries are not able to fully participate in this developing field. What has become known as the digital divide is not just about disparities in ownership and access to digital technologies such as the Internet. Since some people may have access to the Internet but not have the skills to utilize it effectively, a certain degree of digital literacy is required to effectively engage

[2] In his later work, Lanier (2011) shifts from using the term 'cybernetic totalism' to 'digital Maoism', but his underlying argument is unchanged.

with computer technology. Lanier asks us to remember that whilst computer scientists are fantasizing about artificial intelligence and virtual reality, a large part of the population on this planet are concerned with surviving, dealing with the pragmatics of food, health and clean water.

Lanier states that the main tenets of cybernetic totalism include the idea that patterns of information provide the ultimate way to understand reality and that nature and human beings are patterns of information which can be expressed mathematically. In the past the universe was considered to be a giant clock-like mechanism operating in a fixed manner according to Newtonian laws of motion; but more recently, we have seen claims that the universe and the whole of nature are some sort of giant computation. Biologist and mathematician Martin Amos, for instance, claims that the aftermath of the completion of the human genome project in 2003, which was heavily dependent on the use of computerized data processing, has 'transformed biology from a lab-based discipline into a whole new information science' (2006, 2). Going further, Amos states that once we begin to understand 'that computation is not just a human-defined construct but part of the very fabric of existence – only then can we fully appreciate the computational opportunities offered to us by nature' (2006, 13). On the other hand, Denis Noble warns against the universal application of computational principles to areas such as DNA and human life, for according to him there is not a distinction between hardware and software in the human body and there is no list of instructions that the body simply executes. In the following statement Noble also points to the limitations of the analogy between computer programmes and DNA:

> The overall logic of the system may operate in a way that strongly resembles what happens when a program is running a computer, but that does not in itself indicate that there really is such a program. (2006, 53)

In many ways, digital code does appear to be the fundamental building block of life in contemporary culture since it is embedded into the operation of various technological products and services. But we need to be wary of cyber totalism, especially when this becomes a dominant ideological position which discounts other ontological or epistemological viewpoints and explanations about reality.

Presenting a challenge to the tenets of cybernetic totalism, Arthur Kroker and Michael Weinstein take a critical stance towards representations of technology, claiming that the media promotes the concept of a technological utopia. At the same time, they state that there is a 'very deliberate attempt to shut down, silence, and exclude any perspectives critical of techotopia' (2000, 117). Kroker and Weinstein also describe what they call the virtual class, whom they describe as virtual reality manufacturers, developers and computer scientists who have a particular class consciousness which is based on the ideologies of entrepreneurship, intellectual property rights and control of information networks. According to Kroker and Weinstein, the virtual class promotes the idea of social experience as 'prosthetic-after effects: the body becomes a passive archive to be processed, entertained, and stockpiled by the seduction-apertures of the virtual reality complex' (118–119). Whilst I am sympathetic to Kroker and Weinstein's argument, I find the idea of a hegemonic virtual class wielding power and control over technological development and distribution problematic. Kroker and Weinstein assert that the 1990s were 'typified by the rapid decline of the hard ideologies of capitalism and communism, and by the ascendancy of the soft ideology of the virtual class' (135). Of note is that Kroker and Weinstein were writing at the turn of the millennium before the age of social networks and the growth of global companies such as Google, Facebook and Amazon. More recent debates by Eli Pariser (2011) and Lori Andrews (2013) highlight the hegemonic qualities of figures such as Mark Zuckerberg (CEO Facebook), Larry Page and Sergey Brin (founders of Google) and Jeffrey P Bezos (CEO of Amazon). In this regard, there are aspects of Kroker and Weinstein's argument that need to be placed in a more recent context in which the World Wide Web and social networking have become important aspects of contemporary life. A further problem with their argument is that it overlooks complex networks of social relations and the challenges to the ideologies of the virtual class by other fractions within society, such as the GNU General Public License organization, Free Software Foundation (FSF) or the Creative Commons.[3] At the same time, despite these shortcomings, I recognize the importance of the critical debate put forward by Kroker and Weinstein and its contribution to thinking critically about the global capitalist relations and the power that permeates our everyday lives and relationships to technology.

[3] In 1983 Richard Stallman established a free software system. For further information, see GNU General Public License www.gnu.org. Information about the Free Software Movement can be found on their website www.fsf.org. Further information about the Creative Commons, which was established in 2001, can be found at www.creativecommons.org

Discourses

As different representations of virtual reality circulate in contemporary media they become enmeshed within a complex series of power relations and debates. Hegemonic practices can be regarded as an attempt to weave together certain concepts into powerful discourses that can support the agenda of those in power (Howarth, 2000). Political projects such as the commissioning of specialized economic investigations and investment in technology-related research are elements within these powerful discourses and connections. For instance, the development of the Internet, World Wide Web and virtual reality was high on the US political agenda during the late 1980s and early 1990s. In May 1991, a US Senate sub-committee, chaired by former vice-president Al Gore, was held to discuss virtual reality in relation to science, technology and space. A report on the sub-committee produced by Ben Delaney (1991) suggests that the sub-committee discussed competitiveness in relation to Japanese research and the economic advantages of US investment in virtual reality. Meanwhile in the United Kingdom, the Department of Trade and Industry (DTI) and the Economic and Social Research Council (ESRC) were involved in the construction of knowledge about virtual reality during the 1990s. The DTI commissioned the report 'Converging Technologies: Consequences for the New Knowledge Driven Economy' (1998), which made reference to virtual reality as a means of enhancing economic prosperity through its use in education, design, manufacturing and commerce. Similarly, the ESRC developed a £3 million programme on the virtual society from 1997 to 2000 which involved 25 academic institutions.[4] These examples illustrate the ways in which virtual reality technology is related to economic and technological concerns, particularly the shift towards knowledge-based economies.

In Fred Turner's thoughtful and considered analysis of the discourses surrounding digital technology during the 1990s, he notes that the hype surrounding digital technology gave the impression that 'the net would level social hierarchies, distribute and personalize work, and dematerialize communication' (2006, 257). Other discourses, however, focus on the ways in which technology supports the interests of wealthy and powerful members of society. Turner argues that 'despite their usefulness, neither of these perspectives explains just how digital technologies and utopian ideology came together' (258). According to Turner,

[4] For further information about the ESRC Virtual Society initiative, see http://virtualsociety.sbs.ox.ac.uk/intro.html

this gap in the existing literature can be overcome by examining the relationships between ideologies, social groups and practices. Exemplifying this approach, Turner's work emphasizes the complex and multifaceted aspects surrounding technology, social and cultural practices. In doing so he remarks that 'as we study new forms of technologically enabled sociality, we have an opportunity to explore not only online cultures but also the ways in which online collaborations help generate the symbolic and ideological resources out of which all cultures are made' (266). Indeed there are multifarious connections between technology, culture and the practices of everyday life. Therefore, it is important to state that technology alone does not simply cause social or cultural change. Technological determinism is problematic because it is underpinned by a simplistic cause-and-effect relationship which operates in a unidirectional manner. Consequently, this discussion takes a more nuanced approach by considering the ways in which digital technologies have become intermeshed with social and cultural changes. Developments in digital technologies have reverberated in many different aspects of social and economic production and consumption. Technology intersects culture and society in many different ways through innovation, entrepreneurialism, state intervention and consumption practices. As Spanish sociologist Manuel Castells remarks, 'the dilemma of technological determinism is probably a false problem, since technology is society, and society cannot be understood or represented without its technological tools' (2010, 5).

Informational society

Digital technologies contribute to what Castells calls the informational society. Yet the move towards informational societies is not simply a global phenomenon because there are important economic, political and cultural variations which also need to be taken into consideration. Indeed, Castells cautions against a universalizing global approach to informational society and emphasizes the importance of studying technological change in relation to cultural contexts, historical conditions and specific political interventions. He also points out the differences between information society and informational society. The term 'information society' concerns the role of information in society and this could apply to other historical moments, not just the present day, since information plays an important role in the organization, operation and development of different societies. Offering a perspective on this point, Antony Bryant (2010) has remarked that terms such as 'knowledge society' or 'information society' are problematic because they suggest that in the past societies were ignorant and uninformed. Consequently, the term 'informational society' is more useful in this regard because it refers to a situation whereby the production, circulation and consumption of information become a key aspect of social production and power.

However, author and philosopher Sadie Plant presents an account of the darker underside to the relationships between informational societies and the technologies that underpin them. In her scathing critique of the manufacturing of computers in large-scale factories and their highly divided processes of production, she points out that:

> Silicon valley, Silicon Glen, Bangalore, Jakarta, Seoul, and Taipei provide dispersed networks of what U.S. Multi-nationals call 'virtual aliens' to fabricate the wafers, assemble the circuits, set up the keyboards and the screens [and] make the chips that computers on. (1997, 74)

On a similar note, feminist theorist Rosi Braidotti foregrounds the relationships between embodiment and the production of technologies such as virtual reality:

> Capital trades in body fluids: the cheap sweat and blood of the disposable workforce throughout the third world and the wetness of desire of first-world consumers as they commodify their existence into over-saturated stupor. Hyper-reality does not widen out class relations: it just intensifies them. (2002, 245)

The arguments presented by Plant and Braidotti provide an important counterpoint to some of the fantasies of transcendence and wonderment associated with virtual reality by highlighting the importance of embodiment in the production and consumption of such technologies. Put simply, in some cases, the hardware and software that supposedly offer the possibility of transcending the limitations of the physical body through immersion in virtual reality are made in conditions that have serious implications for the health and well-being of the workers that produce them. Furthermore, as Mark Poster (1990) remarks, post-industrial societies do not fundamentally change the operational imperatives of capitalism. So 'service and information workers, like factory workers…remained salaried employees, subject to the capital-labor relation' (1990, 29). Furthermore, Poster points out that information workers may perform 'tedious, repetitive tasks' (29) which have low status and low wages (such as temporary data-entry jobs).

Digital technologies can now be utilized to instantly transfer finance across the globe, a process which impacts our sense of the virtual and the real. Finance may be virtually transferred across continents and this contributes to shifting conceptions of distance and proximity. Additionally, the rapid transfer of finance may seriously impact what in Marxist terms are the forces and relations of production, particularly with respect to the material and embodied aspects of labour. For example, the global financial system facilitates the outsourcing of production and services to different parts of the world to capitalize on low-cost labour. But this process does not just relate to manufacturing; it also impacts service-based employment such as banking and finance. As Zygmunt Bauman observes 'power can move with the speed of the electronic signal – and so the time required for the movement of its essential ingredients has been reduced to instantaneity' (2000, 11). Indeed, we now live in the age that promotes instant access to information, finance and consumption. Yet as Bauman remarks, 'the disembodied labour of the software era no longer ties down capital: it allows capital to be exterritorial, volatile and fickle' (121). In turn, this volatility of capital has an enormous impact on our lives, in terms of employment, affordable housing and access to finance.

Although finance and information may be transferred instantaneously through digital technologies, our bodies are not so flexible. Moreover, technology can be used to extend our physical selves, but at the same time we are anchored in our own body, in the here and now of lived experience which is marked by such factors as class, race and gender. Now, it is possible to extend our reach through communication technologies such as mobile phones or email. Yet in some ways we still like things to remain in the same place because it gives us a sense of security. In everyday life people play different roles – employee, employer, lover, partner, parent, tutor or student – but as Ken Hillis, professor of Media and Technology, remarks, 'our bodies remain with us both as testimony to who we are and as unifying dimension of ourselves within social polyvalency' (1999, xxxi). In agreement with Hillis, the connections between embodiment and identity are not easily discarded.

As we have seen, technology contributes to shifting relationships between space, time and embodiment in a variety of ways. In order to illuminate these shifting relationships, human geographer Nigel Thrift examines 'new kinds of electronic background time-spaces' (2004, 583) and how they incrementally generate new forms of behaviour and being-in-the-world. According to Thrift, the quest to quantify physical and mental labour, which can be traced back to the work of Babbage, has become refined with electronic modes of calculation. Indeed, Thrift coins the term 'qualculation' to encapsulate 'the outlines of a world just coming into existence, one which is based on continuous calculation at each and every point along each and every line of movement' (583). A further dimension to the technological development of the environment was the building of roads, piping and electronic street lighting. In more recent times there has been the construction of fibre optic networks to connect machines and screens.

Thrift's work explores the ways in which such changes become a technological backdrop that has become so encompassing and ubiquitous that we no longer pay attention to them. Yet these changes have profound implications for embodiment and being-in-the-world. For instance, screens are 'ubiquitous interfaces…which demand certain kinds of structured engagement, which are both geophysical and also phenomenological in that they may alter our understandings of space, time and movement' (585). In technologically advanced societies screen-based interfaces such as self-service checkouts at supermarkets or automated teller machines are so familiar that we barely pay attention to them. Yet the underlying programmes that enable us to use different technological objects in such a habitual, automated fashion are generated using numerical processes that can be tracked, measured and monitored. We are living in an environment which intermeshes virtual spaces, mathematical infrastructure and materiality. In this way, it could be contended that we are living in cyberspace but do not consciously recognize this situation. In our time, material spaces and virtual space have become intertwined and embedded in everyday life. Thrift contends that as the logic of this technological infrastructure 'becomes both necessary and general [it will] fade from human perception, becoming a part of the landscape which the body "naturally" adjusts to and which it regards as a normal part of its movement' (586). One of the problems arising from this situation is that if the computational logic of this technological landscape fades from human perception because it is so thoroughly intertwined with the practices of everyday life, then this may limit our ability to challenge the ideologies that underpin these developments. Therefore, it is important to increase our awareness of the social, economic and political ideologies surrounding contemporary media representations of virtual reality and the increasing digitization of various aspects of our daily lives.

Cyberculture

Writing at the turn of the new millennium, John Thornton Caldwell (2000) claimed that there are no particular sets of issues or questions defining the field of cyberstudies. Furthermore, according to Caldwell humanities-based research into electronic technologies tends to regard all technologies in the same way rather than examining the different practices and processes they involve. The point Caldwell raises here is an interesting one since immersion and interactivity in virtual reality environments are often compared to watching a film or reading a book, yet the former potentially provides a greater sense of interactivity than the latter. Humanities-based studies in visual culture are relevant to the present discussion because they utilize a set of theoretical strategies that emphasize the relationships between the viewer, the viewed and the specific cultural contexts in which images are produced and interpreted (Sturken, M. and Cartwright, L. 2001, Barnard, M. 2000).

Studies within the realm of visual culture have produced creative and innovative insights that go beyond traditional art historical methods that focus mainly on the artist as the primary arbiter of meaning. Therefore, this discussion will draw upon the strategies of visual culture to highlight the social and cultural dimensions that intersect representations of virtual reality. At the same time, however, this discussion acknowledges Guy Julier's claim (2006) that visual culture has its limitations because the processes involved in interactivity through virtual reality challenge notions of the viewer and the viewed. In what follows, I attempt to chart some of the ways in which developments in photography and film play a part in the history of virtual reality.

Cinematography

In the 1930s, before the work conducted by the ARC or Sutherland's research into computer graphics and head-mounted displays, Fred Waller, a cinematographer, was working on the development of immersive visual displays for flight simulators. Later, Waller turned his interests to cinema, producing a system known as Cinerama which was developed for the 1939 World Fair in New York. Cinerama films were shot using a three-camera system and then projected on a 180-degree screen, in specially adapted movie theatres (Waller, 1993). The operational aspects of Cinerama are outlined in an application to the United States Patent Office, filed by Fred Waller and Ralph Walker, which was approved in April 1942. The patent application states that the proposed motion picture theatre will produce:

> the effect or illusion that the spectator is actually in and surrounded by the environment depicted. For example, if the scene is laid in a forest, the spectator, within his normal field of vision, will see forest on all sides and overhead, so as to produce the illusion that the spectator is in the forest, rather than merely looking at a picture of a forest covering only a small portion of the field of vision.[5]

[5] It appears that Fred Waller and Ralph Walker's patent application for Cinerama was submitted on 14th September 1937 and granted in April 1942. The serial number for the application is 163, 712.

In addition, the patent outlined the use of binaural sound effects to further increase the illusion of presence in the depicted environment. In September 1952, Waller's film *This Is Cinerama* was first released on Broadway, New York. In an article marking the 50th anniversary of this public showing, Gregg Kimble discusses the impact Cinerama had at the time it was released:

> Running only 13 weeks in one theater in New York, 'This Is Cinerama' was the highest grossing film of 1952. Several more travelogues would follow, climaxed by two dramatic films co-produced with MGM in 1962. (Kimble, 2002)

In the early 1960s another cinematographer, Morton Heilig, patented what he called a Stereoscopic Television Apparatus for Individual Use. Two years later (the same year that Sutherland unveiled Sketchpad) Heilig patented a Sensorama Simulator, which highly engaged the sensory capabilities of the human body, including, sound, smell, touch and sight. Unlike Sutherland, whose research was well funded (partially by the US military), Heilig had continual difficulties funding his work on Sensorama (Rheingold, 1991). Despite funding setbacks, Heilig decided to put together a sort of home-brew simulator system from commercially available parts from electrical and hardware stores. He subsequently found a business partner, formed the Sensorama Company and built a simulator for entertainment purposes that could be used in arcades or cafes. The Sensorama simulator resembled an instant photo booth, complete with seat, screen and curtain. Those who engaged with Sensorama experienced a simulated motorbike ride through New York City, which was produced from filmed images of actual locations. Tactile handlebars provided a sense of touch and the ride was accompanied by such smells as jasmine and hibiscus. The problem was that the simulator was a highly complex piece of equipment and therefore susceptible to breakdowns. Furthermore, whilst the simulator was multi-sensory, it was not interactive. Nonetheless, Heilig's inventions remain a landmark in the development of virtual reality systems, and the work he conducted on olfactory and haptic simulation remains exemplary.

In the 1970s cinematographer Douglas Trumbull produced another variation on immersive cinema. Trumbull developed what he called Showscan, which used 65mm film projected at 60fps to create a heightened sense of realism. In a *New York Times* interview, Trumbull is reported as saying that Showscan 'allows you to break the fourth wall of cinema, to allow performers to relate one-on-one to an audience' (Trumbull, D. in Anderson, J. 2012). What is significant to this discussion is that some of the issues raised by Trumbull's Showscan, such as heightened realism and moving beyond traditional distinctions between the screen and the audience, have much in common with debates surrounding virtual reality. Indeed, virtual reality technology has been promoted as a way of removing the frame surrounding an image and becoming immersed within it. One of the problems with Showscan, however, was that it needed to be shown in purpose-built cinemas. Consequently, this format did not come into widespread use with Hollywood film producers and distributors due to the investment it required.

In recent years developments in digital technologies have started to impact the ways in which films are produced, distributed and consumed. For instance, Michael Allen (2009) discusses the ways in which digital locations can be generated and these may supplement or replace location-based film production. Allen gives the example of the Coliseum scene in *Gladiator* (2000, dir. Ridley Scott) or the film *Sky Captain and the World of Tomorrow* (2004, dir. Kerry Conran), which made extensive use of digital imaging techniques. In addition, films can now be distributed digitally via a central server, which is a huge saving on printing celluloid copies. Developments in 3D digital techniques have also taken place in recent years and the film industry aims to promote these developments to stimulate the dwindling film market which is beset by competition and threats from DVDs, Internet streaming and piracy. Even so, developments in 3D film have not been unanimously accepted as a positive development. In his article, 'Clash of the Wonderlands', Ian Christie (2011) states that sound and picture editor Walter Murch and renowned film critic Roger Ebert have denounced 3D film. Ebert is especially critical of films that are shot in a conventional format and changed to 3D in post-production, such as *The Green Hornet* (2011, dir. Michel Gondry), whereas Walter Murch has concerns about the relationships between 3D film and the sensory motor features of human perception. In particular, Murch contends that with 3D film technology our eyes converge and focus at different points and this unsettles the ways in which our eyes gauge whether objects are near us or further away. Christie argues that humans have adapted in the past to new technologies, such as travelling at high speeds and moving through time and space, though he does concede that in some cases 3D film can lead to headaches and visual discomfort. What is most relevant to this study is that the disparity between the sensory motor features of embodiment and viewing 3D film could impact the degree to which viewers find these films immersive.

Interactivity and computer graphics

Whilst developments in digital imaging techniques such as 3D film can be regarded as attempts to increase the realism and immersive qualities of the cinematic experience, they do not offer opportunities for interactivity. One of the first examples of combining film and interactivity was a project called the Aspen Movie Map, which was produced at the MIT in the late 1970s and early 1980s. The map was produced by undergraduates Peter Clay, Bob Mohl and Michael Naimark and utilized film images and multiple photographs of the streets of Aspen. A special camera system was fitted to the roof of a car and then used to film streets of the town. Later these shots were combined with photographs taken from helicopters and cranes. The movie also included shots produced from photographs of building interiors.

What is significant is that the Aspen Movie Map was interactive as viewers could choose which street they wanted to venture down or what buildings to enter (from a pre-programmed selection). It was even possible to zoom into some parts of the environment, look left or right or even freeze frame. The Aspen Movie Map was funded by the ARPA and was envisaged as a means of helping soldiers become familiar with new territories in combat situations. Although the map was comprised of photographic images, rather than computer-generated simulations, it was one of the first forms of interactive environments and is therefore another important precursor to the development of contemporary virtual reality-based systems. The idea that immersion and interaction with visual images could be used for military and entertainment purposes brings us to the next part of the discussion, which focuses on the development of computer games.

The term 'video game' is often used to refer to games that utilize raster graphics, which are comprised of data that produce a rectangular pixelated grid on a display monitor. However, the term is often used today to refer more generally to various forms of computer games (Kraus, G., 2009). Thomas T. Goldsmith Jr and Estle Ray Mann produced one of the earliest forms of the video game in 1947. This game used an analogue transmitter and a cathode ray tube and involved controlling a dot on screen, as if it was a missile firing at a target. Other early screen-based games include an electronic version of the game *Tic-Tac-Toe* (*Noughts and Crosses*) produced by Douglas Alexander in 1952 and *Tennis for Two* by William Higginbotham in 1958. As Gerard Kraus has indicated, early video games were produced in the context of Cold War research and development, for '[g]overnments on both sides of the Iron Curtain decided to rely on the newly emergent power of computers to simulate scenarios of attack and defence' (2009, 77).

Yet, after the Vietnam War the military funding which permeated academic institutions and virtual reality research in the United States began to be curtailed. US politician Mike Mansfield, who had been particularly critical of US involvement in Vietnam, instigated the Mansfield amendment (1973), which called for a reduction in military funding in academia (Rheingold, 1991). According to Rheingold, the reduction of funding for academic research into virtual reality provided an impetus for some researchers to look for new outlets to pursue their interests, mainly within the emergent computer games industry.

In the late 1970s, Nolan Bushnell, who created the highly successful video game Pong (1972), founded the Atari Company. Atari established a research laboratory, at Sunnyvale in California, to develop video games and explore the possibilities of producing virtual reality-based environments. The Atari lab soon attracted researchers and the company soon became renowned for its video games. One of the researchers at Atari was Jaron Lanier, who used his musical skills to compose sound effects for video games and also began experimenting with computer programming. Lanier had a vision for the development of a user-friendly computer programming language using sounds and images rather than mathematics. Rheingold (1991) recounts how Lanier developed a language called Mandala, which graphically simulated the operations of the computer. So instead of concealing the underlying operations of the code that makes the software operative, Mandala made them visible at the level of the person engaging with the computer so that they could gain some insight into the processes that were occurring.

Also working at the Sunnyvale lab was Thomas Zimmerman, who developed the device known as the data glove (Rheingold, 1991). His idea for the glove stemmed from his interest in the practice of air guitar, which mimics the gestures and intensity of playing an electric guitar along with rock music (Bell, D., 2001). The prototypes for the data glove were made from thin hollow plastic tubes, which were able to conduct light, and by 1982 Zimmerman had patented the design for the glove.

In the 1980s Lanier went on to found the VPL Research, Inc., which sold high-specification technological products such as data gloves to large-scale research laboratories, including NASA. The company also became involved with Abrams Gentile Entertainment (AGE) to develop a commercial version of the data glove for the video games market. These ventures led to collaboration with the toy company Mattel and a product known as the Power Glove, which sold for under $100. In 1987, VPL developed a complete cybersuit (which resembled a diving suit) which had sensors on the knees, neck, ankles and wrists. But unlike the data glove, the suit, which at the time cost circa $50,000, did not translate into a cheaper, widely available commercial product. What is noteworthy about Zimmerman and Lanier's work is that it shows the possibilities of engaging with embodiment and virtual reality, through various interfaces such as cybersuits and data gloves. Whilst the products they produced were expensive for individual consumers, their work in this area has been particularly influential within industrial- and work-based settings.

By the 1990s, in the United States, there was also a growth in the crossover between military simulations and computer games. Working with commercial software companies, former members of the military began to use their expertise and knowledge to develop games and this was often regarded as giving such games added kudos. Moreover, as Douglas Kellner points out, the genre of military-based computer games is now well established and continues to be highly popular:

> In 2002, the US military developed a highly popular and critically acclaimed computer game, freely available to anyone online for downloading and playing upon registration with the US army. Promoted as 'The Official Army Game' it allows the user to participate in simulated military action. (Kellner, 2003, 10)

America's Army is a first-person shooter (FPS) game with pre-programmed rules and levels of interaction, which shapes the gaming experience and the meanings arising from it. But David B. Neiborg (2004) points out that the game can be interpreted as a propaganda tool for promoting the US army. The use of gaming as a recruitment method is an effective marketing strategy in terms of the target demographic for the US army because the game is likely to be particularly appealing to young men who have been raised with computer games and are familiar with them. Furthermore, more recent versions of the game appear on a sophisticated website which includes promotional videos, links to YouTube, user-created missions and an open invitation to become a Facebook fan, a process seemingly designed to draw upon social networking as a means of extending the marketing drive for military recruitment.

In terms of the academic scholarship on gaming, attention has also been given to massively multiple online role-playing games (MMORPGs). These are large-scale online environments where users cooperate and sometimes compete to achieve their objectives. Within this burgeoning academic field on gaming attention has been given to online communication and the ways in which co-operation and trust can be built between different players (Castronova, E., 2005, 2006). In particular, researchers have explored the ways in which social rules, norms and practices are extended or subverted in these online games. One of the well-known massively multiple online role-playing games (MMORPG) is *World of Warcraft*, which was established in 2004 by Blizzard Entertainment. The number of people who play *World of Warcraft* is now equivalent to the population of some countries such as Sweden and Bolivia (Corneliussen, H. G. and Walker Rettberg, J., 2008). And *World of Warcraft* exceeds the usual definitions of a computer game since it has multiple levels and no definite beginning or end. New quests and goals are continually added to this world and completing a particular quest may take a considerable amount of time and effort.

Within *Warcraft* there is an in-game auction house for trading items which generates a thriving network for economic exchange. Commenting on the economic dimensions of *Warcraft*, Scott Rettburg contends that it is a simulation of life in the capitalist workplace and 'reinforces the values of Western market-driven economies' (2008, 20). In particular, he argues that *Warcraft* 'offers players a capitalist fairytale in which anyone who works hard and strives enough can rise through society's ranks and acquire great wealth' (20). Rettburg points out that *Warcraft* encourages players to work with or lead others and accumulate personal resources. Players are also subject to performance management through the ranking of their avatars according to such factors as strength, stamina, agility and the level of the game they have reached. They are also ranked according to the items or resources they have required, such as weapons or tools. In this way, the operational features of *Warcraft* resemble performance reviews and target-driven measurement in the contemporary workplace. According to Rettburg, *Warcraft* players can also earn gold or increase their social standing by engaging in various tasks such as 'picking herbs, mining precious metals, or tailoring garments' (27). Often, the tasks that need to be undertaken to earn these credits, virtual gold or other resources are repetitive. What is interesting is that players engage in this form of repetitive labour online during their leisure time to earn these extra credits or virtual items. But there are also instances of gamers in China who are paid to play *World of Warcraft* to accumulate in-game resources which they can auction 'in house' as a means of trading currencies with Western players. For example, if you are new to *Warcraft* and do not want to earn the necessary credits to go up a level in the game, you can buy virtual gold or tokens from other game players via the in-game auction. As Rettburg remarks, there is a mismatch between the medieval guilds, quests and iconography of the online world and the sophisticated electronic technology it employs. So 'while much of the *World of Warcraft* seems modelled on a very analog, medieval/feudal culture, both the game's mail system and auction system are based on more contemporary electronic mail systems and markets' (29).

Whilst studies of MMO games or MMORPGs can yield important insights into collective behaviour from a social science perspective, it is equally important to critically analyse other types of computer games and the sort of experiences they offer. Of note is that players do not watch a game unfold; rather, they co-create the game-playing experience through the choices they make and their interaction with the images and situations depicted on screen. Corliss (2011) also points out that there is often a distinct temporality to such experiences. For example, the form and structure of the game encourages players to be future orientated, to reach a goal, reward or complete a mission.

Another factor to consider is the ways in which user-generated content contributes to the production and circulation of computer games. In this context, user-generated content is known as mods (or modifications) to a game. Mods are often produced by experienced players who have the ability to re-write the code and change the operating parameters of a game. In this way these players are customizing and creating new game-playing experiences that go beyond those envisaged by the companies that produced them. Yet, as Toby Miller (2006) has pointed out, sometimes game mods are used to further extend the interest in a particular title and can lead to further revenues for the official game producers. Furthermore, experienced game players may be motivated to provide their coding expertise for free, modifying existing games in the hope that this will lead in the future to employment within the games industry.

In sum, cyberculture has been shown to be an important arena in relation to the concept of virtual reality, especially in relation to the aesthetics of representational realism and the experience of immersion. This discussion has traced the ways in which developments in cinema such as Cinerama, Sensorama and Showscan served as a means of enhancing the cinematic experience in various ways. Cinerama and Showscan offered increased realism through the use of widescreen, panoramic technology. In this way, these cinematic experiences attempted to immerse audiences within the film image. Although Sensorama was designed for individual use, rather than for a large audience, it aimed to provide a multi-sensory immersive experience which can be regarded as a precursor to computer-generated virtual environments.

Furthermore, the Aspen Movie Map was also an important development on the way towards the sort of interactive imagery that can be found in computer games. The military dimension to computer games has also been discussed since this continues to be an important area as US military personnel utilize computer simulations for training purposes. In turn, military-based computer games such as the *Call of Duty* series continue to be extremely popular with players and have generated huge revenues for the companies that produce them, often eclipsing the box office returns of Hollywood films.

As discussed, in recent years MMORPGs have also become incredibly popular with millions of players. For instance, we have seen that *World of Warcraft* is based upon a digital infrastructure that enables players from different geographical locations to come together in a shared virtual world. Within the digital realm of *Warcraft*, players can explore different roles and enter into different sorts of social relations. In this discussion we have also seen the ways in which the virtual world of these games intermeshes with real-world economics, as virtual items are traded for actual currencies such as the US dollar. By and large, this discussion has indicated that these developments in cinema and gaming point to the ways in which immersive and interactive imagery have become familiar aspects of contemporary culture.

This chapter has surveyed the ways in which computer-generated virtual reality environments are underpinned by developments that took place during the nineteenth century, particularly in relation to mechanized algorithmic processing. As noted in this chapter, ideologies surrounding the relationship between humans and machines were connected to the development of mechanized forms of calculation. In the nineteenth century, the production of the mechanized loom had a major impact on the physical work of weavers, whereby small-scale production was displaced by large-scale, factory-based textile production. This shift from small-scale production to industrialized modes of employment brought about major social changes in employment and living conditions. We also saw that Babbage's work on mechanized forms of calculation impacted the mental labour performed by human beings. But this chapter indicated that mechanized calculation also transformed the relationships between humans and machines and the way they were compared and evaluated.

Taking these historical strands into consideration it is possible to challenge the idea that virtual reality is a radically new technology, for virtual reality environments are generated by algorithmic calculations which can be traced to technological developments and the ideologies surrounding mental and physical labour in the nineteenth century. Furthermore, contemporary debates surrounding the relationships between digital technology and the displacement of human labour can be regarded in the context of earlier debates about the displacement of human physical and mental labour by machines in the nineteenth century.

This chapter has also discussed the ways in which the algorithmic processing conducted by electronic computers was subsequently used to develop visual imagery in the context of post–Second World War military research and development. One of the driving forces for these developments was the requirement to produce military simulators to train pilots and soldiers for combat situations. However, running parallel to military-based computer applications was the development of computer games for entertainment purposes.

The move towards industrial production, the displacement of human modes of calculation and the development of computer-generated imagery are also relevant to the growth of the computer games industry. In the past 30 years, the de-regulation of financial markets and increasing global interdependence have generated massive shifts in terms of the production and consumption of digital technologies, especially products such as home computers, tablets and smart phones. Capital can flow almost instantaneously across continents through the flickering figures on computer screens. However, as Plant and Braidotti have argued, it is important to remember that the glossy advertising images for digital technologies are underpinned by large-scale industrial production often in the Far East.

Our bodies are not as fluid as digital modes of finance and the workers who produce the digital technologies that are promoted within contemporary media are often engaged in highly repetitive work and long working hours. Computer-generated environments have also been shown to be the site of other forms of contemporary production and exchange. This chapter has indicated, for example, that in some cases Chinese game players of *World of Warcraft* engage in repetitive tasks within this environment to generate items that can be traded for currencies such as the dollar, and similar practices also take place in the popular virtual world of *Second Life*.

Overall the discussion has indicated that it is important to examine the historical, social and cultural factors surrounding virtual reality. In this discussion we have also considered the ways in which debates around virtual reality have been positioned politically, economically and culturally, from Al Gore's championing of the information superhighway to Timothy Leary's celebration of transcendence from embodiment through immersion in virtual worlds. Going further, this chapter explored the ways in which the information superhighway was associated with the growth of the informational society whereby computing processes intersect our everyday lives. Equally though, Leary's proclamations of wonderment and transcendence from embodiment will be shown in subsequent chapters to remain an important feature of representations of virtual reality in contemporary media.

2

Virtually Real and Really Virtual

This chapter will outline the ways in which the virtual and the real are not necessarily discrete categories; rather, it will make evident that the virtual and the real are mutually inflecting. In other words, some things are deemed more virtual than real, whilst others are regarded as more real than virtual. From the outset it is important to acknowledge that the relationships between the virtual and the real have a long and complex history. Throughout history and in different cultures and societies alternative realities have been explored in a variety of ways through incantations, meditation, prayer, dream states, hallucinogenic plants and drugs. In visual culture, the use of perspective, anamorphoses and Trompe L'Oeil paintings produced the illusion of three dimensions on a two-dimensional surface. Devices such as the Camera Obscura, Zoetrope and Magic Lantern have also aimed at producing a convincing illusion of reality. Offering a view on this issue, founding editor of *Wired* magazine Kevin Kelly states, 'We have always made miniature worlds – of dolls, novels, and plays – but these new worlds (think *Sim City* or *Ultima Online*), so expertly crafted on a matrix of silicon, are of a kinetic complexity never before possible' (1999, 388). In this statement Kelly acknowledges the historical dimensions of illusion in visual culture and also emphasizes the ways in which computer-generated imagery offers a new aesthetic. Going further, he suggests that computer-generated worlds can also provide compelling interactive sensory experiences that go beyond previous forms of visual culture. But whilst Kelly's pronouncement has a ring of optimism, we shall see that Jean Baudrillard's work provides a more critical and dystopian view of simulation.

Baudrillard argues that simulation is particularly prolific in contemporary culture and shifts our sense of the virtual and the real. According to Baudrillard, we 'cannot even imagine how much the virtual – as though running ahead of us – has already transformed all the representations we have of the world' (2002, 108). Taking these views into consideration, this chapter critically analyses the representation of virtual reality technology in contemporary culture. As Erik Davis points out, virtual reality is 'a concept that exceeds mere gadgetry and all its inevitable bugs and breakdowns. The concept is absolute simulation' (1999, 247). To begin this critical analysis of the virtual and the real, there will be a brief sketch of the intellectual tradition and context in which Baudrillard's work was produced.

Baudrillard was part of the French intellectual scene which emerged in the late 1960s along with Guy Debord (1931–1994), Jacques Lacan (1901–1981), Claude Levi-Strauss (1908–2009) and Michel Foucault (1926–1984). These intellectuals could be considered to be part of an emergent but heterogeneous intellectual group. Their work could also be regarded as challenging and building upon the previous generation of European thought based upon the works of Karl Marx (1818–1883), the sociologist Emile Durkheim (1858–1917) and the existentialist movement. Initially, Baudrillard worked within a sociological intellectual tradition, emphasizing the material aspects of social production. Yet, as his theoretical work developed, he moved towards arguments relating to symbolic signification. In *The Consumer Society*, for example, Baudrillard states that production and consumption are entangled (1998, 33). From this perspective the capitalist aspects of material social production are intertwined with economic markets that consist of pulsating data streams within global networks of exchange. Arguably, in the contemporary global age, the relationships between these pulsating data streams that are graphically represented as pixels on computer screens and the forces and relations involved in manufacturing and distribution are becoming increasingly fraught. Whilst the nodal points of post-industrial economies are centred in cities such as London, New York and Tokyo, other parts of the world are still centred upon industrial modes of manufacturing.

Virtual reality technology is part of a complex chain of production, distribution and consumption within global markets. For instance, there is a complex division of labour and expertise between hardware manufacturers, software companies and entertainment and leisure industries. The forces and relations of the production of virtual reality involve actual embodied labour which is located in factories, laboratories, warehouses, shops as well as sites of consumption such as leisure centres and theme parks. The physical location of workers across the globe is also mediated by and through technological systems such as computer-based assembly lines or the data control mechanisms of distribution. Consequently, there is no clear-cut distinction or separation between the electronic economic realm and the myriad of power relations between employers and workers in the virtual reality industry. Computer memory chips might be made in Singapore, whilst a display screen is produced in Hong Kong and the software for the virtual environment could be written by teams in California or Bangalore. Once the virtual environment software and hardware is produced, it has to be transported, merchandized, sold and purchased (Plant, 1997).

Talking about the complex relations surrounding technology, embodiment and the forces and relations of production, Donna Haraway states:

> Our best machines are made of sunshine; they are all light and clean because they are nothing but signals, electromagnetic waves, a section of a spectrum, and these machines are eminently portable – mobile – a matter of immense human pain in Detroit and Singapore. People are nowhere near so fluid, being both material and opaque. (1991, 153)

Since there is a complex division of labour within material production and the supply chain for consumer goods, even those who claim to use computer technology with relative ease may not necessarily know where these products are made, or how they work. So even if we claim to be computer literate we are actually working with a readymade product. Those who use virtual reality technology may not know (or care) how it is produced, or how the software is programmed. Yet virtual reality is not unique in this regard as many other products and services are produced and consumed without understanding the complex divisions of labour and skills they involve.

A key difference between Marxism and Baudrillard's later work is the concept of agency. Marxism as a theoretical strategy aims to highlight inequalities within the forces and relations of production in order to bring about social change. But Baudrillard argues that consumer society is driven by symbolic systems in which commodities are used as a means of differentiation and stratification. Whilst individuals strive to differentiate themselves through the acquisition of commodities, Baudrillard claims that the actual forces and relations of production are unable to meet this demand. Indeed, consumer culture aims to stimulate rather than satisfy our desires since this is a way of perpetuating the profit-driven dimension of the capitalist system. However, one of the drawbacks of Baudrillard's arguments in *Consumer Society* is the claim that agency is circumscribed by the operational parameters of signifying systems.

Code

As virtual reality environments are generated by computer code it is important to examine the ways in which these codes operate at a practical and metaphorical level. Focusing on computer code is important from a semiotic perspective since it raises questions about the production of meaning via signifiers and what they signify. A key question in this regard is, will the computer codes that underpin virtual reality environments institute a new series of relationships between signs, meaning and referents? To explicate these issues further the discussion turns to the operational parameters of symbolic systems of signification.

Mathematical systems of notation and calculation are signifying systems which are fundamental to the operation of virtual reality computer systems. Mathematical codes are derived from earlier Hindu and Arabic systems, which are thought to have spread to the West along trading routes. The Egyptians, Greeks and Romans used the abacus to perform calculations whereas the Hindus used a symbolic form of notation to calculate and record results. Sadie Plant (1997) makes the point that the Christian Church deified the concept of the number one by associating it with the existence of one God, one universe and one creator. When mathematical notation was introduced, the zero was regarded as threatening to the system because it differed from the abacus, which did not rely on the zero to perform calculations. Plant goes on to say that the zero was assimilated to become absence, nothing and non-being. These ideas are important because it is the relationships between different electrical pulses codified as zeros and ones that form the operating systems for virtual reality environments. Furthermore, the development of computer code raises some interesting issues in relation to establishing a material referent that exists beyond the symbolic system.

Some of the earliest forms of language were systems of iconic marks that made a connection between the visual appearance of a thing, event or person and written forms. However, in Western culture as systems of written language developed the connection between visual appearance and the meaning of words became more complex. Languages are now thought of as systems involving syntax, combination and structure. The development of graphic systems of notation and speech became the vehicle for a shift in human consciousness because it enabled words to become conscious. Indeed, from the perspective of Western culture, language can be an object of reflection. Commenting further on this issue, Barbara Johnstone points out:

> One way in which language and thought might be related is that the possibility of human cognition and human experience might depend on the fact that humans are language-users. 'Thought', 'language' and 'being human' might in other words all be part of a single activity. (2002, 32)

There are also cultural differences in terms of language as signifying systems. For example, Yasuo Yuasa (2005) states that in Western languages meaning operates retrospectively after a sentence is spoken. Japanese Kanji characters, however, represent an idea or meaningful unit rather than a particular syllable or sound. The operational aspects of signifying systems are relevant to the representation of virtual reality in contemporary culture through linguistic discourses and imagery. On this point, Jaron Lanier claims that virtual reality could offer new forms of symbolic communication to enable different ways of thinking as well as alternative forms of creative experience. According to Lanier, virtual reality could be 'a new type of conversation' (Lanier quoted in Frenkel, 1995, 48). He explains how this conversation might differ from other conventional forms of communication. For example, he says that he could suddenly compose the sentence 'you and I have turned into crystalline, spiral-shaped swans, twisting above the Seas of Titan, dancing the samba' (1995, 49). Here Lanier is making the point that with language it is possible to construct various idiosyncratic realities from the rules of the system in which it operates. Similarly, it might be possible to create virtual environments where people could communicate and interact through images and sounds rather than words. After outlining some of the issues relating to codes, images and symbolic systems of signification, I will now move on to critically analyse Baudrillard's idea of simulation and hyperreality in more detail.

Simulation and hyperreality

In *Simulations* (1983), Baudrillard discusses a quotation from Ecclesiastes which states that simulacra, which are signs that refer to other signs, rather than a referent beyond the symbolic system, do not simply hide the truth. Upon investigation, Jonathan Stuart Boulter (2001) states that the quotation in Baudrillard's work does not actually appear in Ecclesiastes, but it does point to the difficulties in defining the real, the copy and the original. Baudrillard states that simulacra are not manipulative because they do not simply mask or hide the truth. Instead, simulacra are subversive because they challenge the notion of truth or reality. Gary Genosko (1994), William Merrin (2001) and Michael Camille (2003) have all pointed out that the word 'simulacrum' can be traced back to Plato's dialogues, where it was used to refer to phantasm or semblance. The concept of the simulacrum concerns the relationships between image and reality. In Plato's philosophical schema the simulacrum was a false likeness which leaves the real/unreal binary intact. As William Merrin reminds us, in Western culture:

> The image has always been conceived of as powerful, as possessing a remarkable hold over the hearts and minds of humanity – as having the capacity to assume for us the force of that which it represents, threatening in the process the very distinction of original and image. (2001, 88)

Within this historical context the image was interpreted as something that can captivate the viewer who can then become enthralled by its power, for the power of the image seems to come from its ability to acquire the properties of that which it represents. Furthermore, Merrin says that within Western culture there have been attempts to counter the threat to reality which is posed by simulacra. One way of neutralizing the power of simulacra has been to institute the concept of the original and the copy. In this way the copy becomes a reflection or mediated version of the real, thereby bolstering the notion of a prior reality.

In his discussion of simulation Baudrillard also mentions a fable by Jorge Luis Borges (1899–1986) about a map which is so detailed that it covers the territory it represents. Baudrillard's point is that in the past there was a theoretical gap between the map and the territory it represents, but in contemporary culture the simulacrum threatens this relationship. In this way the map in the fable could be regarded as analogous to the ways in which the symbolic order overlays our experiences and understanding of concepts such as reality and the virtual. Baudrillard claims:

> We live as if inside Borges's fable of the map and the territory; in this story nothing is left but pieces of the map scattered throughout the empty space of the territory. Except that we must turn the tale upside down: today there is nothing left but a map (the virtual abstraction of the territory), and on this map some fragments of the real are still floating and drifting. (2000, 63)

In Boulter's discussion of Baudrillard's reference to Borges, he tells us that it comes from *Del Rigor en la Cienca (Of Exactitude in Science)*, which is 'about how the Real is displaced by its representation' (2001, 355). In addition, Boulter remarks that Borges's text is a transcription of a portion of a text by J.A.

Suárez Miranda, called *Viajes de Varones Prudentes* (1658). Reflecting on this intertexuality, Boulter contends that 'part of the effect of Borges' writing is to call into question the very notion of origins' (2001, 356), and this is one of the reasons why it is central to Baudrillard's concerns about simulation and reality.

Baudrillard's argument regarding simulation is underpinned by a fourfold structure which he calls the successive phases of the image (1983). The first phase begins with images that are thought to be a reflection of a basic reality. The second phase of the image occurs when the image masks and perverts reality. But by the third phase the image masks the absence of reality. Then in the fourth phase the image 'bears no relation to any reality whatever: it is its own pure simulacrum' (1983, 11). Baudrillard also attempts to map these stages onto different economic conditions from feudal times, to mercantile capitalism to advanced global capitalism. Yet it is unclear if these phases of the image can be neatly mapped out in this way. Instead, it seems that the changing aspects of imagery require further consideration since Baudrillard focuses on the ways in which one form of imagery succeeds another and becomes the dominant aesthetic, rather than exploring how these phases might overlap. So whilst the simulacra may be a dominant form of the image in contemporary culture, the other three phases of the image still remain in place.

According to Baudrillard, Borges's text is an example of the second order of simulacra which masks and perverts a basic reality. But taking a different stance, Boulter says that Borges's text can be interpreted according to all the four phases of the image that Baudrillard proposes. He contends that the text 'reflects an originary ontology, masks that ontology, masks the absence of that ontology (or attempts to), and, as pure modernist text, is its own simulated ontology: the text's "real" reality' (2001, 362). What I find particularly interesting about Baudrillard's model of the succession of the image is that it could also be used as a means of providing multiple interpretations of images in virtual reality environments. For instance, it is possible to interpret a virtual environment as a reflection of a reality. In this regard, an architectural model of a building or a flight simulator could be thought of as an attempt to accurately reflect a building or flight conditions in the real world. Secondly, military combat scenarios, which often become translated into entertainment games, might be regarded as masking and debasing the atrocities of war. Cynics might also argue that the virtual and the real are both symbolic constructions that have no ultimate foundation. Alternatively, the virtual could become indistinguishable from the real, thereby becoming hyperreal.

However, in the following statement Baudrillard insists that there are fundamental differences between representation and simulation:

> Representation stems from the principle of the equivalence of the sign and of the real (even if this equivalence is utopian, it is a fundamental axiom). Simulation, on the contrary, stems from the utopia of the principles of equivalence, from the radical negation of the sign as value, from the sign as the revision and death sentence of every reference. (1994, 6)

Extending this line of argument, Baudrillard suggests that simulation involves the construction of its own referential ground, a process which he terms 'the hyperreal' (1994, 1). So what we find in Baudrillard's fourfold schema is that signifying systems are misleading because they generate the notion of reality as something that is outside of the system through the very operational processes of the system itself. Amplifying this point, Gary Genosko remarks:

> In Baudrillard's terms, every time there is signification, there is lying, for the reason that what is real is an effect of the sign, and thus, every referent is an alibi: signification simulates reference to a real state because no real state corresponds to the sign. (1994, 41)

Baudrillard's central argument is that in contemporary culture simulacra attempt to absorb the real by becoming equivalent to reality. For Baudrillard states, 'The image cannot imagine the real any longer, because it has become the real. It can no longer transcend reality, transfigure it, nor dream it, because it has become its own virtual reality' (1998, 4). In other words, if the image becomes so close to reality that there is no relationship of resemblance, the process is short-circuited; the image becomes reality. The hyperreal is too visible and over-signified, whereby the dialogue between the viewer and the viewed also becomes strained. As Merrin puts it:

> At such a level of obviousness no relationship is possible: there is no passion, no investment or belief – everything is hyperrealised before us and our only response is stupefied acceptance. (2001, 100)

As Baudrillard's work develops, another stage is added to his procession of successive images which he calls the fractal, whereby the notion of the referent changes to that of the combinatorial. Meaning is produced by the combination of different signs within a signifying system. This idea of the fractal is also associated with the proliferation of signs in contemporary culture, particularly the speed in which they are produced, circulated and exchanged. To explicate this process,

Baudrillard creates an analogy between the fractal stage of signification and cancer, whereby the cells proliferate to such a degree that they become dysfunctional. Likewise, in Baudrillard's fractal stage signs of reality proliferate so that they begin to act like a virus which unsettles notions of truth and reality. In this way, signs do not make the world intelligible; instead, they make it more unintelligible.

A recurring theme in Baudrillard's work is that virtual reality does not just apply to a specific technology; rather, it is an apt description of life in a contemporary media-saturated world. According to Baudrillard, in contemporary culture our everyday lives have become so permeated by media images that our sense of reality has become occluded by hyperreality:

> We don't need digital gloves or a digital suit. As we are moving around in the world as in a synthetic image. We have swallowed our microphones and headsets, producing intense interference effects, due to the short-circuit of life and its technical diffusion. (1999, 19)

But whilst the aesthetics of photorealism are a prominent feature of popular culture, fine art has gone in a rather different direction. As Andrew Darley (2000) points out, the aesthetics of realism in fine art declined during the twentieth century whilst this became a major aspect of popular culture through film, television and more recently computer games. On a similar note, Joanna Buick states that 'commercial enterprises and technical wizards are racing towards ideals of faster computation, perfect simulation and life like simulation. In art things are somewhat different' (2002, 109). Darley also claims that digital imagery is not necessarily conducive to semiotic analysis, or the textual uncovering of meaning, because it is more concerned with stimulating the senses. For Darley computer-generated images 'tend towards pure diversion, consisting of forms that are immediate and ephemeral in their effect' (2000, 6). Michael Heim also challenges conventional ideas about semiotic analysis in his consideration of digital imagery. Heim (1995) argues that the computer graphics that construct virtual environments are not signs that re-present that which is not present. Instead, he asserts that virtual entities do not 'present again' something which is absent or 'already present somewhere else' (1995, 70). This means that simulated entities are taken to be real in a fundamentally different way to the realism in painting because 'the symbol becomes reality' (1995, 70). Similarly, Nicholas Mirzoeff states that 'unlike photography and film which attest to the necessary presence of some exterior reality, the pixelated image reminds us of its necessary artificiality and absence' (1999, 30). Taking these remarks into consideration, we can see that it is important to give critical attention to the aesthetics of photorealism and how this impacts our sense of the virtual and the real.

Military simulation

Critical debates about realism and computer-generated imagery have also been raised with respect to military-based simulations. For example, in an interview with Kara Platoni, Captain Pete Huntley from the US Marine Head Quarters in Quantico commented on the use of simulation techniques within a military context. According to Huntley, 'the goal is to make the simulation be almost indistinguishable from actual wartime needs so the soldiers have no way of knowing if they're in a real conflict or a simulated exercise' (Platoni, 1999, 30). What this quest for photorealism indicates is that crossovers can occur between computer games and military-based virtual reality simulations. Kara Platoni states that 'thanks to massive improvements in graphics technology over the last few years, the video game software industry has produced shooting, flying, and fighting games that look so real they can be used for actual combat training' (1999, 27). Platoni goes on to say that in 1997 the US Marines made a few changes to the commercially available game *Doom* (Id Software) so that it could be used for military training purposes. Subsequently, the revised version of *Doom* was made available as a free download on the Internet by the military.

The ways in which military or combat-based computer games represent life, death and violence are an important consideration, particularly in relation to our understanding of the virtual and the real. Steven Poole (2000) points out that killing in game scenarios is often rewarded and symbolized by the accrual of more points. Killing can be programmed into computer games so that some characters in a game are killed in a different way to others. For example, when a player kills an enemy they may appear to regenerate, come back to life and attack them again. Moreover, in some games players are given multiple lives, and as Poole argues, these features of game playing have the potential to shift the ways in which we think about life and death:

> In part this resembles the brutal calculus of war, where a human life, normally the definition of total value in peacetime, is arithmetised as being worth, say one hundredth of the value of taking the next ridge. (Poole, 2000, 68)

Poole's argument about life and death in gaming is also pertinent to the use of simulation techniques in actual combat situations. For instance, the 1991

Gulf War raised a series of concerns about simulation techniques and how they impacted the real world. The Gulf War was not the first war to be televised, but it was the first time that images were supposedly broadcast in 'real time' across the world's major television networks. Paul Patton says that the Gulf War also involved specialists who were responsible for representing the war in particular ways via the mass media:

> As a result, what we saw was for the most part a 'clean' war, with lots of pictures of weaponry, including the amazing footage from the nose-cameras of 'smart bombs', and relatively few images of human casualties, none from the Allied forces. (1999 121)

Patton's comments reveal the ways in which representations of the war were skewed through media representations, which obscured the actual death and destruction that was taking place. Therefore, the US involvement in the Gulf War was not just about the control of weapons, places or actual human bodies, but it also involved controlling images and information and shaping public opinion.

Benjamin Woolley also discusses the use of computer simulations in the Gulf War of 1993 and how simulators were used to transmit data from Norman Schwarkopf's command centre to Washington D.C. Woolley states:

> Targets were not real locations but map coordinates displayed on a VDU, troop movements were formations of pixels in computer-enhanced false-colour satellite images. From a postmodern perspective, the entire war, at least at the level where anyone could make sense of it, was just a pattern on a screen. (1993, 93)

The testimonies of the US soldiers fighting in the Gulf War that appear in Woolley's work also reveal the ways in which events are understood within the context of media representations. A US pilot said that the events that he was involved with were 'exactly like the movies' (1993, 191). In addition, the worldwide circulation of televised images of the Gulf War also produced a feedback loop between images and the events they purportedly recorded. Writing about these events in the Gulf, Baudrillard comments that the war was speculative 'to the extent that we do not see the real event that it could be or signify' (1995, 29). According to Paul Patton, who translated Baudrillard's book, the purpose of this approach was to take the logic of a media-based war to its ultimate extremes in order to reveal its absurdity.

Journalist Nick Davies (2009) has also investigated media representations of the Gulf War. According to Davies the US military produced a media strategy that involved 'a $12 million campaign orchestrated by PR specialists from Hill and Knowlton, which used misinformation from Kuwati exiles to provide a series of fabricated stories' (2009, 221). Furthermore, Davies found that the representation of what was happening in the Gulf became increasingly complex and that nothing was exactly what it seemed. For example, Davies reports that what appeared to be a grass-roots organization called 'Empower Peace' was established by the US State Department. In addition, the Iraq Crisis Bulletin, which claimed to present independent news, was actually created by Voice of America. What Davies's work shows is that uncovering or untangling what is real and what is fabricated is becoming increasingly complex, particularly in terms of media representations of war.

Davies also discusses the ways in which the production of the news in a global context that operates 24 hours a day, 7 days a week also impacts how the news is represented, particularly in terms of accuracy and validity. He recounts that due to the increasing rapidity of deadlines, newspapers and television companies often regurgitate material from other news outlets without checking the validity of their sources. Davies's argument was highlighted by an incident in which a news bulletin on the channel TV2 in Denmark used screenshots from Ubisoft's computer game *Assassin's Creed* to represent a news item about Damascus in Syria in March 2013. In the reports that followed it was claimed that the researcher for the channel obtained the image of Damascus from YouTube and incorporated this into the news bulletin without checking its authenticity. In this instance, the virtual photorealistic screenshot from a computer game was used to represent the real.[1]

Taking a different view, George Friedman claims that simulation can be used as a deterrent by showing the destruction and loss of life that could occur, if a war actually took place. Friedman reports that the Wright Patterson Air Force Base used simulations of battle scenarios to convince 'the Serbians that the United States, could in fact, defeat them and to convince them that the U.S. was prepared to defeat them' (1997, 62). Then again, Lanier is critical of Friedman's argument about using simulation as a deterrent because he claims that there is 'an extraordinary gadget lust' driving the military (1997, 62). Like Lanier, I am not entirely convinced by the deterrent argument. Lanier goes to the core of the ethical debate by asking why war takes place and suggests how simulation might be used in alternative ways. Indeed, Lanier wants to utilize virtual reality in a creative way to produce 'problems that are harder than warfare to take up

[1] The use of the image from *Assassin's Creed* in the TV2 news bulletin was reported in the Huffington Post 11th March 2013 (see http://wwwhuffingtonpost.co.uk accessed 26th April 2013).

people's time' (1997, 62). Lanier's proposal of using virtual reality as a creative tool to avoid war is a noble one, but one wonders how practical and effective it might be when placed next to the multimillion-dollar research budgets and rhetoric of the US military. On the other hand, there are virtual environments that attempt to simulate complex and multifaceted situations in order to explore different solutions, such as *Darfur Is Dying*, which is a simulation of the experiences of refugees in Sudan, or *PeaceMaker*™, which is based on the conflicts between Israel and Palestine. Then again, it is debatable whether virtual reality as a creative tool to avoid violence is what people actually want since violence is represented in popular computer games in the military genre such as the *Call of Duty*® (Activision) or the *Medal of Honor* series (Electronic Arts).

Big Data

More recently, new debates have emerged with respect to modelling, simulation and the management of catastrophic events through the accumulation of large data sets (known as Big Data). At present, different data sets such as weather patterns or geological changes can be brought together in order to help predict the likelihood of a hurricane or tsunami. The work of Israeli artist Eyal Gever provides interesting insight into the creative potential of Big Data. Gever spends up to two years modelling virtual worlds which are based on data acquired about catastrophic events or phenomenon. He then simulates the catastrophe from different perspectives and uses a 3D printer to transform the virtual environment into physical reality. This technique is used to produce such works as *Tsunami Crashing* (2011/2012) and *Street Blast* (2011/2012).[2] In this way, Big Data provides the starting point for a modelling process in which the virtual (the potential) helps to predict what could become real.

The use of Big Data as a means of modelling scenarios such as the movement of weather patterns, geological changes or population patterns indicates the ways in which the digital technology of computer code forms a backdrop to the mapping, measuring and understanding of space (virtual or otherwise). In the next section we will see further examples of the ways in which Baudrillard's conception of simulation is interwoven with real places and spaces.

[2] Further information about Eyal Gever's work can be found at http://www.eyalgever.com

Simulated spaces

Baudrillard contends that simulation bolsters the idea of real social spaces. For example, Baudrillard claims:

> Disneyland is presented as imaginary in order to make us believe that the rest is real, when in fact all of Los Angeles and the America surrounding it are no longer real, but of the order of the hyperreal and of simulation. (1983, 25)

He also observed this simulation process through the prism of the Parisian shopping centre Parly 2. Providing critical commentary on Parly 2, Baudrillard states that it is 'adorned with fountains, artificial trees, pavilions and benches', and in this way it appears to be 'wholly exempt from changes of season or bad weather'. Yet it is also 'an exceptional system of climate control, requiring 13 kilometres of air-conditioning ducts' which gives the impression of 'perpetual springtime' (1998, 29). Going further, Baudrillard claims that Parly 2 is a model of how contemporary urban life might develop:

> The drugstore writ large in the form of the shopping centre, the city of the future, is the sublimate of all real life, of all objective social life, in which not only work and money disappear, but also the seasons, those distant vestiges of a cycle which has at last also been homogenized. (1998, 29)

Baudrillard may have a point since in the past 25 years there has been a growth in simulated social spaces. In the United Kingdom, the Sno Zone chain has leisure centres at Milton Keynes, Glasgow and Castleford which offer consumers the simulated experience of skiing, in 'perfect' skiing conditions all year round. Indeed, the Sno Zone website note proclaims: 'It has snowed every day at Sno Zone since opening and each slope is covered in 1700 tonnes of real snow.'[3] One line of argument is that some people may actually prefer simulation to the uncertainty, danger and differing climatic conditions of the 'real' world. In *Travels in Hyperreality* (1986), for instance, Umberto Eco talks about his visit to Disneyland and how simulation could be preferable to reality:

> Disneyland not only produces illusion, but – in confessing it – stimulates the desire for it: A real crocodile can be found in the zoo, and as a rule it is dozing or hiding, but Disneyland tells us that faked nature corresponds much more to our daydream demands. (1986, 44)

[3] http://www.snozoneuk.com

Similarly, Sno Zone seems to provide a sanitized and controlled environment for skiing and this could be part of its appeal.

Another example of the ways in which simulation intersects actual places is Huis Ten Bosch in Nagasaki Prefecture, Japan, which is a simulation of a seventeenth-century Dutch town. There are some historical dimensions to the resort since Nagasaki was the only Japanese port open to International Trade during the Edo period (1603–1868) and to commercial interactions with the Dutch. The Huis Ten Bosch resort was built in the early 1990s from land that was reclaimed from the sea and has around 150 buildings which are replicas of buildings in Holland. Although these buildings are replicas, the relationships between the real and the copy, or the authentic and inauthentic, are more complex than initially appears. In The Hague, Huis Ten Bosch is one of the residences of the Dutch Royal Family. This Royal residence was built in 1645, but it was severely damaged during the Second World War and has been extensively rebuilt and refurbished. So strictly speaking it is hard to locate an original, or authentic, Huis Ten Bosch even in Holland.

The Huis Ten Bosch resort uses specific Dutch motifs such as tulips, windmills and canals to simulate a sense of visiting Holland. So, arguably in this resort visitors can enjoy the surface play of Dutch history without having to go to Holland. So why pay air fares or succumb to jet lag? Instead, visitors can enjoy the simulated Dutch experience in this Japanese resort which features an array of contemporary conveniences such as luxurious hotels, a wellness spa and restaurants serving Italian, French, Chinese and Japanese cuisine.

What is interesting about Huis Ten Bosch is that visitors are not just viewing simulated images because they are embodied and immersed in actual social spaces. As such, visitors can stroll along canal pathways and smell fragrant tulips and dahlias. In this way, visitors appear to have a real experience of simulated and idealized Dutch history. Remarking on the authenticity of the simulated experience, the Huis Ten Bosch theme park website states:

> Kinderdijk is the first scenic spot to greet you as you enter Huis Ten Bosch. You'll feel like you've actually gone to the rural Dutch countryside as you walk among the meandering canal and windmills towering into the sky. Kinderdijk abounds with tulips in spring and other flowers from each and every season.[4]

If visitors are able to stroll along canal pathways and enjoy looking at windmills and tulips, as described in the promotional material for this resort, does it matter

[4] Huis Ten Bosch theme park website (http://english.huistenbosch.co.jp).

if these experiences take place in Japan rather than Holland? Although it could be argued that Huis Ten Bosch may offer an idealized version of Dutch History, to what extent is historical accuracy possible, necessary or even desirable? For life in Holland during the seventeenth century was an amalgamation of many different social, religious, political and economic factors that cannot be simply replicated through a visit to a leisure resort.

In contrast to seventeenth-century Dutch life, which was underpinned by early forms of mercantile capitalism, the Huis Ten Bosch resort is enmeshed in contemporary forms of global capitalism. For example, the resort website extols visitors to 'experience the Extreme Floods of the Netherlands', through the Horizon Adventure Plus experience. Here visitors 'feel like they're about to be engulfed as mist, lightning, waves, rain and deluges of water approach with frightening realism, all brought to life by special equipment'. In addition, 'this production uses diorama to realistically recreate the Netherland's scenery' through a combination of 'powerful video images' and even '800 tons of actual fresh water'. This dramatic Horizon Adventure Plus is an experience that is enmeshed within capitalist economic relations, which equates time with money since it lasts 20 minutes and costs approximately 600 yen (which is equivalent to about £5.00).

To summarize this section, by examining the Huis Ten Bosch theme park in Japan it is possible to explore the varying dimensions of the relationships between the copy and the original. What I find most intriguing is the idea that in some cases it is possible to have embodied sensory experiences within a simulated place. For example, Huis Ten Bosch is subject to seasonal change, which adds a particular nuance to the visitor experience. Visitors can have a real embodied experience of strolling along canals and viewing flowers that are currently in season. On the other hand, visitors can equally have powerful experiences such as Horizon Adventure, which provides a thrilling albeit limited simulated experience of extreme weather conditions. But there are also important distinctions to be made between actually experiencing the life-threatening conditions of an actual flood and purchasing a simulated thrilling experience at a resort, or in a leisure park.

Theoretical challenges

Whilst Baudrillard's theoretical framework provides a useful entry point for the discussion of the virtual and the real, such arguments have come under criticism from other writers such as Douglas Kellner (1989) and Christopher Norris

(1996). Norris suggests that Baudrillard's work exemplifies postmodern theory, particularly the idea that there is a 'sceptical mistrust with regard to all truth-claims' (182). Consequently, instead of providing a theoretical ground which is based on 'normative standards' or 'veridicial knowledge' there is an emphasis on what is 'currently and contingently 'good in the way of belief'' (182). What Norris seems to be saying here is that Baudrillard's work is based on an anything goes attitude towards knowledge and truth. From this perspective truth is not investigated and unearthed; rather it rests upon the presentation of a persuasive argument. In what appears to be a critique of Baudrillard's work, Norris states:

> We have now lived on (so the argument goes) into an epoch of all-embracing 'simulation' or 'hyperreality', a world where – quite simply – there is no possibility of distinguishing truth from falsehood, fact from fiction, or the order of genuine (veridical) knowledge from its various 'ideological' fantasy-substitutes. (1996, 183)

Meanwhile, Kellner contends that Baudrillard's work is premised upon a fetishism of signs that 'ascribes more reality to signs than things' (1989, 180). In response to these criticisms, it is worthwhile pointing out that Baudrillard's work changes over the years from the Marxist and Structuralist leanings of his earlier work to his later work which emphasizes signification and hyperreality. Upon close inspection Baudrillard's work does offer critical arguments that are based on distinguishing truth from falsehood. For example, during an interview with Nicholas Zurbrugg, Baudrillard says that he offers 'a very critical account of technology and of technology's impact on the world' (Zurbrugg, N., 1999, 38). In this regard Baudrillard's statement indicates that his work is underpinned by a process of evaluation and at some level it does distinguish fact from fiction. What is particularly relevant to this study is that Baudrillard's work does provide a lucid description of the operating functions of symbolic systems of signification which are a fundamental feature of virtual reality environments.

If Baudrillard's work provides a valid description of contemporary society and simulation, then what theoretical strategies can be used to challenge the proliferation of simulacra? Furthermore, how might these strategies answer the criticisms that Norris and Kellner raise against Baudrillard's work? In *Consumer Society*, Baudrillard compares and contrasts 'primitive' and contemporary cultures as a means of constructing the symbolic as a theoretical strategy with which to challenge simulacra. According to Baudrillard, 'primitive' societies were based on symbolic exchange and he uses this term in a specific way. For Baudrillard, the symbolic is a reciprocal system based on kinship ties and the establishment of social obligations. In contrast, he states that contemporary consumer culture is based upon simulacra and hyperreality which are not grounded in a reciprocal exchange system. Illustrating these arguments, Genosko states that 'the symbolic is Baudrillard's revolutionary anthropological antidote to the political economy of the sign' (1994, 5). However, one of the difficulties with Baudrillard's theoretical engagement with 'primitive' societies is that he presents wide-sweeping statements which are not supported by reference to any particular historical or ethnographic studies of such societies.

Baudrillard also uses the concept of seduction as a theoretical strategy. Seduction in this context involves enticement, for Baudrillard states:

> Seduction is not that which is opposed to production. It is that which seduces production – just as absence is not that which is opposed to presence, but that which seduces presence, as evil is not that which is opposed to good, but seduces good, as the feminine is not opposed to the masculine, but seduces the masculine. (1998, 58)

In this way Baudrillard opens up the possibility that the relationships between different elements within a signifying system could be based on attraction rather than opposition. Baudrillard shores up this argument by referring to signifying systems of the past:

> In ancient cosmogony, the elements of water, earth, fire and air were not the distinctive elements of a system of classification, but attracting elements, seducing one another: water seduced fire, fire seducing water. (1998, 58–9)

So instead of searching for a referent, or recovering reality, Baudrillard devises a strategy that involves leading signification astray by seduction.

Baudrillard's theoretical strategies changed over the years and in later works he emphasized the notion of pushing signifying systems to an extreme so that they implode. He argues that operational systems become dysfunctional at some point. For example, too much information can result in information overload. According to Baudrillard:

> There is a strong possibility, verging on a certainty, that systems will be undone by their own systematicity. This is true not only for technical structures but for human ones as well. The more these political, social, economic systems advance toward their own perfection, the more they deconstruct themselves. (2000, 78)

Arguably, this idea of signifying systems imploding is relevant to the global financial crisis of 2008. This occurred at a time when computerized systems were processing financial transactions across continents almost simultaneously. Many factors contributed to the global financial downturn, including human greed, the deregulation of global financial markets and the use of mathematical modelling in which finance became increasingly abstracted from material conditions. For instance, physicists were employed on Wall Street and introduced systems that modelled the financial markets in abstract and speculative ways through quants and derivatives (Weatherall, 2013). Yet, at some point these systems became overextended as overinflated property prices and loans were issued to people who did not have the economic means to keep up repayments. In this way, the global financial crisis was not just a technical structural problem; it was a human problem as well. Out of this situation, the virtual world of the global financial market of quants and derivatives became painfully real when people's homes were foreclosed, companies went into administration and jobs were lost.

The material consequences that have occurred in the aftermath of the 2008 global financial crisis are also relevant to Baudrillard's contention that the forces and relations of ontological reality go beyond symbolic signification. According to Baudrillard, there is something that is not entirely caught up by signification because it is 'indecipherable' (1988, 63). What exceeds signification is a void; it has no meaning because it cannot be represented or understood symbolically. Yet Baudrillard asserts that this void is something that is kept at bay through signification because the loss of meaning is a terrifying possibility:

> And to keep it at bay, we have to realize the world, give it force of reality, make it exist and signify at all costs, take from it its secret, arbitrary, accidental character, rid it of appearances and extract its meaning, divert it from all predestination and restore to its end and its maximum efficiency, wrest it from its form and deliver it up to its formula. (1996, 16)

Reality can be explained through different symbolic formulas, through scientific analysis, experimental testing and debate. Yet whilst description and analysis allow the construction of knowledge about reality, there are still limits to this process. If the operational aspects of symbolic systems of signification can no longer be judged according to idealistic notions of the real, Baudrillard suggests that such systems can never be completely totalizing. There must be a gap that enables meaning to seep through the procession of simulacra. Baudrillard's point is that 'there is no place for both the world and its double' (1999, 27). So here is the suggestion that hyperreality has not been perfected; simulacra are not quite equivalent to reality. For as Baudrillard remarks:

> Nothing is identical to itself. We are never identical to ourselves, except perhaps, in sleep and in death. Language itself never signifies what it means; it always signifies something else, through this very irreducible, ontological absence from itself. (2000, 70–71)

Furthermore, computer viruses, breakdowns and system anomalies prevent symbolic systems from becoming totalized. In other words, there is still a space for reality to be perceived. There is still a gap between the image and what it represents. If there were no difference between simulacra and reality, if the hyperreal was perfected, then there would not be a critical space for such debates about simulation and reality. As Merrin (2001) remarks, if Baudrillard's work provides us with an accurate means of understanding contemporary culture because simulacra has swallowed reality, then this would mean an end to his theoretical framework. There would no longer be any distance between his theoretical representation and that which it represents. The theory would have taken on the qualities of that which it represents, in this case, simulacra. On the other hand, if the concept of the uprising of simulacra is denounced, then paradoxically this actually is a means of supporting Baudrillard's work.

As discussed, Baudrillard's work highlights some of the issues surrounding contemporary media representations of virtual reality and how this relates to shifting ideas about what constitutes the virtual and the real. Baudrillard's work concerns the ways in which consumer society operates through seduction, stimulating our desires. Yet the desire for commodities is never satiated. The point is desire keeps the economic system circulating, perpetuating consumption and profit. Technological commodities are an important aspect of life in the twenty-first century as consumers strive for the latest technological products and services. Virtual reality is also part of this process through the consumption of immersive experiences at theme parks, leisure centres and museums. Somewhat hidden beneath the surface of commodity culture are the forces and relations of production that make them possible. And from this perspective, virtual reality technology is part of a global nexus of production and the uneven distribution of wealth, from the low-paid assembly line workers to the corporate directors of multimillion-dollar media conglomerates.

This chapter has outlined Baudrillard's claims that in contemporary culture representation has been usurped by simulation. Consequently, signs are no longer upheld by a referent outside of the system itself. But even though money flows in the form of electronic symbols within a global financial system, this does have physical and material consequences. After the global financial crisis of 2008, homes have been foreclosed, companies have folded and in some cases governments have shored up the banking system to prevent the collapse of the financial system itself.

A further and more ethically troubling dimension to simulation can be found in the context of contemporary warfare. In this context, military targets may appear as pixels on a computer display unit and resemble the aesthetics of computer games, but firing weapons can have devastating effects upon human life. People are killed or injured and towns and cities are destroyed. In this chapter, we have seen that Lanier suggests that virtual reality can be used as a creative catalyst to divert aggressive tendencies into more productive activities than war. In agreement with Lanier it would be beneficial for aggression to be channelled into creative pursuits although it seems difficult to envisage how this might be achieved on a wide-scale basis. Additionally, the US military is a major economic juggernaut in terms of research and investment into virtual reality; so it is unlikely that this military agenda for virtual reality would be abated in the foreseeable future.

Nevertheless, it is important to think ethically about the implications arising from regarding 'the enemy' as pixels on a computer screen and the ways in which this can make it easier to kill others, at a distance and with no hesitation. Military personnel who are involved in combat situations may no longer have to be physically present as they engage with the 'enemy'. Instead, military personnel may find themselves in a situation whereby they press buttons on a control pad and watch events unfold on a computerized display. In this way death and destruction may become increasingly abstract and more reminiscent of computer game-playing experiences.

However, a plausible criterion for the existence of physical reality is that it 'kicks back' (Deutsch, 1997); in other words, it produces a result that requires adequate explanation. David Deutsch suggests that reality can be thought of as a series of forces and relations that can be tested to produce results. This does not preclude the possibility of error as new methods and means of testing physical reality are developed. Whilst reality kicks back, the totality of the ontological realm remains elusive. It seems that Baudrillard's notion of system anomalies can be read in a positive light because they provide a space for meaningful debates about reality. Structures and systems can never be complete because something always exceeds them. Absolute presence or the absolute simulation of reality is impossible in an ontological sense and Baudrillard seems to regard this as a positive thing because without such differences meaning is lost. As Baudrillard says, this elusiveness is what prevents simulation from occluding reality by becoming its double. In conclusion, it seems that simulation can be a useful model or tool in a variety of settings, but it will not result in complete hyperreality. What counts as the virtual and the real will continue to be represented in contemporary media and subject to ongoing debate.

3

Technological Intimacy and Social Estrangement in *Strange Days*

Strange Days, a novella by James Cameron (1995), is set on the eve of the millennium in Los Angeles. The central feature of *Strange Days* is a fictional device called a SQUID (superconducting quantum interference device), which enables one person's experience to be extracted from their cerebral cortex and then played back by someone else. Although the use of cortical recordings for personal purposes is illegal, *Strange Days* also suggests that they have a legitimate function within the criminal justice system because they provide testimonial verisimilitude and can be used as evidence in court. Whilst *Strange Days* is not overtly about computer-generated environments, it does play a significant part in my study of contemporary media representations of virtual reality because it represents cortical recordings as a way of offering extremely potent immersive experiences. On this basis, I will insist that *Strange Days* presents us with several concerns about technologically mediated experience and immersion that require further attention.

James Cameron started writing his novella in the mid-1980s, but the film version, which was directed by Kathryn Bigelow, was not released until 1995.[1] This chapter will refer to the *Strange Days* novella, but on the whole it concentrates on the ways in which the film represents technologically mediated experiences. One of the reasons for focusing more on the film rather than the novella is that the film vividly represents immersive experiences in a visual way, often using point-of-view (POV) camerawork in order to place the spectator in the position of someone who is using cortical recordings themselves. What is intriguing about this use of POV camerawork is that it seems to invite spectators to form an intimate connection with the characters in the film and their experiences.

[1] The screenplay for *Strange Days* was written by James Cameron and Jay Cocks.

At the same time the intimacy that appears to be afforded by this vision of cortical recordings is juxtaposed within a wider framework of social tension, alienation and estrangement.

In this chapter I will focus on three main themes. Firstly, I consider *Strange Days* as a film noir thriller that represents illicit and in some cases macabre, technologically mediated experiences within a backdrop of heightened social tensions, economic inequalities and estrangement. Secondly, I investigate the ways in which cortical recordings are represented as having greater verisimilitude and authenticity than other aspects of visual culture such as film or television. On this basis, the film suggests that cortical recordings can be used within the criminal justice system as a legitimate form of testimonial evidence. Finally, I consider whether the film is a cautionary tale which points to the problems arising from privatized immersive pleasures. From this perspective *Strange Days* draws our attention to the importance of embodied experiences that are grounded in co-present spatial and temporal relations. Overall, the main argument put forward in this chapter is that cortical recordings cannot fully capture and recreate another person's experiences. However, by engaging with *Strange Days* it is possible to consider the implications arising from technologically mediated experiences, such as the context in which they have arisen and how they can be interpreted and understood. To date, *Strange Days* has not received sustained critical attention and appears to have disappeared under the scholarly radar, though there are some significant pieces written on the film by Laura Rascaroli (1997), Jay David Boulter (2000), Mark Berrettini (2002) and Brian Carr (2002). But as this chapter will indicate *Strange Days* opens up a series of important concerns around technologically mediated experiences that remain relevant today and deserve further scholarly attention.[2]

In 1995, when the film version of *Strange Days* was released media representations of virtual reality were particularly nascent. During this time virtual reality was often represented as a revolutionary and unprecedented immersive experience in popular literature and film. In his discussion of *Strange Days*, Jay David Boulter observes that cortical recordings can be thought of as 'a fanciful extrapolation of contemporary virtual reality, with its goal of unmediated visual experience' (2000, 63). Therefore, *Strange Days* can

[2] Bigelow's oeuvre is often studied in relation to spectatorship. For example, the film *Blue Steel* has received critical attention in relation to spectatorship and the representation of gender (Tasker, 1993; Powell, 1994; Rascaroli, 1997), whilst *The Hurt Locker* (2008) has been studied in relation to the representation of war (Lauzen, 2011).

be contextualized in relation to other representations of virtual reality and the Internet in Hollywood film during the 1990s. In particular, *Strange Days* can be interpreted alongside *Total Recall* (1990 dir. Paul Verhoeven) and *The Lawnmower Man* (1992, dir. Brett Leonard). What we find in these films are a set of concerns about how immersive experiences in virtual reality impact subjectivity, temporality and memory. For instance, in *Total Recall* the protagonist, Douglas Quaid (Arnold Schwarzenegger), purchases a compelling immersive experience of a trip to Mars from a company called Rekall. But the narrative drive of the film concerns the destabilization of identity and experience, so viewers are unsure whether they are watching the immersive experience Quaid has purchased from Rekall or if what they are seeing supposedly represents reality. Meanwhile, in *The Lawnmower Man* immersive experiences are represented as so compelling that the main character Jobe jettisons his physical body for a post-embodied life in virtual reality.

Furthermore, these popular films can also be regarded alongside the discourses of transformation and exuberance that surrounded virtual reality technologies and the rhetoric of the Internet as an information superhighway or new frontier that were prevalent during this period. This is not to say that such films presented an entirely favourable view of technology. Instead, these films were the site of complex and contradictory positions regarding technological change and how this impacts upon embodiment and social relations. In this sense, *Strange Days* can be regarded as very much a product of a particular moment in time and is embedded within a set of complex social, cultural and economic relations. Even so, whilst the concept of using head-mounted display devices to access immersive virtual worlds has not come into widespread usage, the idea of immersive visual experience continues to be an important aspect of contemporary culture.

When Cameron was writing his novella he was setting out a vision of the millennium, but this is now part of recent history. Therefore, a key question is to what extent are the themes of the novella and the film still relevant today? In this regard, *Strange Days* will be shown to be a cautionary tale of what could happen if privatized mediated experiences start to usurp face-to-face relationships with others, both in our immediate locale and beyond. These are interesting points to consider at a time when technologically mediated communicative experiences are now becoming commonplace through widespread usage of the World Wide Web and the use of social networking sites. Moreover, immersive experiences have become pervasive and commercially accessible through the use of gaming consoles such as Microsoft X-Box, Nintendo Wii™ and the Sony PlayStation®.

Another more recent development is augmented reality, whereby small, lightweight devices (such as glasses or smart phones) can be used to provide digital recording capabilities and graphical informational displays that overlay our experience of the world around us. For instance, the US company Google are developing a device that records what the wearer sees, and these visual experiences can also be shared instantly through links to social networking technology.[3] Therefore, it seems timely to engage with the representation of technological intimacy and social estrangement in *Strange Days* in more detail.

Estrangement

Strange Days is set in the backdrop of Los Angeles at the turn of the millennium at a time of civil unrest. Arising out of this urban malaise is an increased emphasis on privatized experience to avoid the potential dangers of interaction with other people who also inhabit the city. But with this increased emphasis on privatized experience and a reduction in face-to-face relationships social antagonisms start to rise further. Setting the film in Los Angeles seems significant since this is a city that is regarded as the economic and cultural centre of the film industry in the United States. Indeed, Los Angeles is regarded as a city that produces popular cinematic fantasies that entertain, inspire and engage a mass audience. Therefore, Los Angeles is a city that is associated with compelling visual imagery, desire and seduction. Yet, Cameron's novella provides a different vision of Los Angeles since the city is portrayed at a time when 'racial and class tensions are higher than ever [and] the city seems constantly on the verge of chaos and martial law' (13). Surveillance and private security are represented as growing industries, and the rich attempt to protect themselves from the violence and chaos of the city streets in gated communities. This representation of Los Angeles also chimes with Mike Davis's (1990) study of class, racial divisions and urban space in his influential book *The City of Quartz*.[4]

A sense of social estrangement is evoked in an early establishing sequence in which the protagonist Lenny Nero (Ralph Fiennes) is driving along the

[3] http://www.google.com/glass/start/

[4] Today, from the perspective of the twenty-first century, gated communities, surveillance and the uneven distribution of wealth in society continue to be important concerns. In February 2012, for example, unarmed teenager Trayvon Martin was killed by George Zimmerman, a member of a neighbourhood watch in a gated community in Florida. This event has sparked debates about the criminal justice system, racial and social inequalities.

city streets in the early evening. The dark city streets are lit with Christmas decorations and banners comprising of multi-cultural images promoting 'World Wide – Faces of the World'. But the situation on the city streets bears little resemblance to these idealized poster images of multi-culturalism and festivity. Instead, we see two young men standing outside a shop brandishing machine guns and further along the street there are armoured tanks. The extra-diegetic sound of the wailing of sirens and police helicopters also adds to the sense of social unrest that pervades the city.

Lenny has dropped down the social scale from his respectable role as a member of the Los Angeles Police Department's Vice Squad to a dealer who sells illicit cortical recordings. Outlining Lenny's character, in his novella Cameron writes:

> Lenny is an operator. A fast talking salesman, scam artist. Hustler. A dealer. He is a creature of the moment…To him the city is like a big coral reef …a big food chain. Alive and dynamic. A place where a fast fish can avoid being eaten. (19)

Lenny is a character who mainly uses other people because they are useful to him or as a means of self-gratification by capitalizing on selling lurid recordings to his clients that aim for affect rather than intellectual engagement. Lenny circumnavigates bars, clubs and hotels looking for affluent customers who want the playback hardware and software that will provide them with the experience and thrill of other people's lives. As Lenny drives along the city streets in a flashy Gold Mercedes, he listens to a radio talk show in which the presenter asks one caller if they are looking forward to the New Year. The caller replies: 'What's the point nothing changes New Year's Day … the economy sucks … fifth grade kids are shooting each other … whole thing sucks right?' The points made by the caller just at the turn of the millennium could also be made in relation to more recent events, such as the aftermath of the global economic crisis of 2008 and anxieties concerning the shooting atrocities that took place at Columbine (1999), Virginia Tech (2007) and Sandy Hook Elementary School (2012). In fact, the issues raised by the radio listener in this sequence still seem relevant to contemporary social concerns and recent events.

As Lenny listens to reports of social unrest he is cocooned in his car and appears to be physically separated from what is happening on the streets around him. Whilst passing by scenes of urban unrest, he stuffs food into his mouth and conducts a rapid-fire business transaction on his cellular telephone whilst the tensions around him continue to flare.[5] This physical separation of Lenny from the social interaction and conflict which is taking place on the city streets also sets the tone for the theme of social estrangement through privatized experiences within the film narrative. Commenting on social interaction, Richard Sennett remarks that the city is 'a human settlement in which strangers are likely to meet' (1978, 39). But when we traverse city streets we may barely acknowledge one another and may even devise a range of strategies for avoiding or limiting our encounters with strangers. On a similar note, Zygmunt Bauman states:

> Strangers meet in a fashion that befits strangers; a meeting of strangers is unlike the meeting of kin, friends, or acquaintances – it is, by comparison a mis-meeting. In the meeting of strangers there is no picking up at the point where the last encounter stopped, no filling in on the interim trials and tribulations or joys and delights, no shared recollections: nothing to fall back on and to go by in the course of the present encounter. (2000, 95)

In other words, we do not have a personal shared history with strangers, nor are we planning a shared future with them. In some instances, this lack of connection with strangers can also give rise to estrangement, fear and a lack of understanding. In turn, such estrangement can contribute towards social unrest since different inhabitants of the same city may start to think in antagonistic ways towards one another because they no longer cross paths on a regular basis. As a result, the idea of connecting to others through privatized mediated experiences may actually start to appear attractive, for in the privacy and apparent safety of our own homes we can choose who we want to connect to and interact with at a time when it is convenient for us. Indeed, Michael Bull's (2004) seminal study of mobile music technology highlighted the ways in which people utilized MP3 players to create their own auditory bubbles, in which they customized urban spaces through listening to their private playlists. In this way public spaces became privatized and more pleasurable through the use of mobile music technology. However, *Strange Days* presents us with a fictional scenario in which affluent members of society are forgoing social interaction in public places for privatized immersive playback recordings of other people's experiences. Notably, Sherry Turkle (2011) has argued that whilst technology does not cause social estrangement in a deterministic way, technologies such as

[5] In the mid-1990s owning a cellular telephone was a status symbol since these devices had not yet come into widespread use.

social networking do make it easier for us to communicate at a distance and to limit risky and uncertain face-to-face contact.

Testimonial verisimilitude

The idea of technologically mediated experiences is vividly captured in the opening sequence of *Strange Days*, through the representation of a robbery which utilizes POV camerawork. The scene begins with a close-up shot of a single eye which quickly cuts to first-person perspective shots of the back of a car. The image judders and jerks, highlighting the frantic motion of the car as it speeds along the road. Through the use of first-person perspective camerawork the viewer is invited to experience the robbery as if they were there. But the scene ends on a disturbing note as one of the robbers is cornered by the police and jumps to their death from a high-rise building.[6] Explicating this process in greater detail, Laura Rascaroli (1997) remarks that what we are seeing through this use of POV is 'an intentional gaze which is actually charged with a triple presence: that of the film's body, that of the character's body, and finally that of the spectator's body' (234). In other words, the use of POV shots can be interpreted as an attempt to mimic subjective perception and first-person experience. But even though we appear to be witnessing the scene through one of the character's eyes, this is a cinematic construction.

In the following scene we see Lenny and realize that he has been watching the cortical recording of the robbery through trodes attached to his forehead. After viewing the recording, he throws down the trodes and tells supplier Tick (Richard Edson) that he does not deal in snuff material for ethical reasons. Furthermore, Lenny is annoyed because this recording will need editing before he can sell it on to his customers. What we can gather from these remarks is that there is an underground, illicit market and a criminal network for the trading of cortical recordings as a commercial product. At the same time, Lenny's comments about editing undermine the idea that cortical recordings provide unadulterated access to reality.

Yet as Brian Carr (2002) remarks, cortical recordings are represented as going further than other forms of visual culture because a person 'cannot only see, but be in the place of another, a spectatorial transportation that puts one into direct access with an alternate life world; unlike in television or the cinema'

[6] The use of frenzied camera movements and chase sequences are common themes within Bigelow's work. In this sense, the opening scenes from *Strange Days* are reminiscent of Bieglow's earlier film *Point Break* (1991).

(192). From this standpoint cortical recordings also appear to go further than previous attempts at verisimilitude in the tradition of Western painting and in photography because they are extracted straight from the cerebral cortex. As Vivian Sobchack reminds us, initially photographic images were associated with 'testimonial verisimilitude' because they appeared to objectively record reality (1994, 90). Similarly, the ways in which cortical recordings are extracted from the cerebral cortex gives the impression that they objectively record reality. Mark Berretini (2002) explains that within the context of *Strange Days* cortical recordings were initially developed from wiretapping techniques and are used by the intelligence community and law enforcement agencies. In this way, cortical recordings can be regarded as a dramatic exaggeration of existing surveillance technologies. In the following passage from his novella, Cameron also outlines the underlying premise of authenticity, objectivity and cortical recordings stating that '[t]he courts argued over the admissibility of cortical recordings as evidence. But it was decided that they were more reliable than video and audio recordings' (14). This statement suggests that initially there were doubts about the authenticity of these recordings. However, Cameron adds that '[s]o far the technology does not exist to manipulate or falsify the cortical recordings' (14). Therefore 'what you see is what happened' (14). If the technology does not yet exist to manipulate or falsify these recordings what will happen when this does become possible?

As discussed, *Strange Days* raises important issues about the authenticity of cortical recordings, especially within the criminal justice system. These issues are highlighted in a sequence involving the murder of black rap singer Jeriko One (Glen Plummer) by two white police officers, Burton Steckler (Vincent D'Onofrio) and Dwayne Engelman (William Fichtner). Cameron's novella places the murder within the backdrop of economic and organizational tensions within the criminal justice system. It is a time when '[t]he court system is collapsing under the weight of rampant street crime, cops are understaffed and underfunded, frustrated and mean, alienated from the communities they protect' (16–17). The sequence begins with Jeriko One, a wealthy and successful rap artist, driving an expensive convertible car in downtown Los Angeles accompanied by a small entourage. The group are in a party mood since it is New Year's Eve; they are laughing and listening to loud rap music when they are abruptly pulled over by the police for no apparent reason. After pulling over the car, Steckler and Engelman recognize Jeriko One and begin to intimidate him. In turn, Jeriko One taunts the police officers, telling them that he makes more money in a day than they do in a year. After Jeriko One makes a series of disparaging remarks about the police, Steckler snaps and shoots him at close range.

One of Jeriko One's female companions Iris (Brigitte Bako), who was wearing a SQUID device, flees the scene and is pursued by the police. Iris manages to evade the police officers but has now become their primary target because she has a cortical recording of the Jeriko shooting, which could be used as evidence in court. Considering that *Strange Days* was released in the mid-1990s, the recording of the Jeriko One incident can be interpreted in the light of the Rodney King incident of 1991 when video footage showed his repeated beating by members of the California Highway Patrol. This video footage which was shot by amateur cameraman George Holliday was subsequently circulated by an array of different news channels, generating intense public debate and social unrest.

Strange Days also contrasts the realism and authenticity of cortical recordings with media representations of Jeriko One's murder. For instance, Lenny learns about Jeriko One's death through watching a television news bulletin that provides misleading information since it claims that his murder appears to be gang related. At a time when large global media conglomerates provide a major proportion of the production and dissemination of our daily news bulletins, there are increasing concerns about the truth values of the representations we encounter on television or read in the newspapers (Davies, 2009). Taking the importance of truth values into consideration, the idea of testimonial verisimilitude through cortical recordings appears to be an attractive proposition, for the idea of cortical recordings suggests that we could gain first-hand experience of world events.

Researchers have already started to explore the possibility of creating compelling immersive news. For instance, the research conducted by de la Peña et al. (2010) aims to produce an immersive experience of the news in which a participant 'can enter the story in one of several forms: as oneself, a visitor gaining first-hand access to a virtual version of the location where the story is occurring, or through the perspective of a character depicted in the news story' (2010, 292). De la Peña et al.'s research involved generating a scenario based on the interrogation of detainees at Guantánamo Bay Prison (circa 2002–2003). The scenario was based on actual information about these events which was acquired through the Freedom of Information Act. The idea was that when a person was immersed within a virtual environment they would have the impression of being the detainee. Sound recordings based on actors reading the transcripts of actual interrogations also aimed to heighten the realism of the experience. The researchers subsequently analysed the responses from those who participated in the project. One participant was reported as saying, 'I really felt like I was a prisoner in Iraq or some … war place and I was being interrogated' (298). Summarizing their findings, the researchers claim that 'immersive journalism does not aim solely to present the facts, but rather the opportunity to experience the facts' (299).

Yet it is important to question whether news audiences would like the opportunity to experience the facts through immersive journalism. Do viewers really want to feel as if they are there, witnessing or experiencing some of the terrible atrocities that we see on television news bulletins? Furthermore, would viewers become overwhelmed or simply immobilized by the testimonial verisimilitude of such immersive news experiences? Or conversely, would audiences start to become immune or less emotionally responsive to such news experiences as they become more commonplace? These questions are also pertinent to the representation of cortical recordings in *Strange Days*. For to what extent do the characters in the film empathize with others through immersing themselves in cortical recordings? Furthermore, to what extent can the representation of cortical recordings invite the cinematic audience to question their care and concern for others?

Thrilling experiences

Strange Days is set at a time when 'reality shows and amateur video shows dominate TV programming. It is the age of scopophilia, voyeurism, and vicarious living' (Cameron 1995, 18). Scott Bukatman's discussion of *Videodrome* (dir. David Cronenberg, 1982) is relevant here since it features a set of striking connections between the multi-sensory aspects of embodiment and television viewing that is also pertinent to *Strange Days*. In *Videodrome*, the character Professor O'Blivion makes the following pronouncement:

> The television is the retina of the mind's eye. Therefore the television screen is part of the physical structure of the brain. Therefore whatever appears on the television screen emerges as raw experience for those who watch it.

The protagonist in *Videodrome*, Max Renn, is the leader of Civic Television, a channel that specializes in broadcasting unsavoury material including grotesque and sadistic acts. Max seeks to boost the channel's ratings by moving to something edgier by purchasing a show that broadcasts snuff material. However, a company called Spectacular Optical develops a signal for the Videodrome channel that produces a tumour in the viewer that causes hallucinations. These hallucinations are then played back to the viewer so that they are caught in a fatal, mediated experience. The idea that television could provide such raw and disturbing experiences is similar to the representation of cortical recordings in *Strange Days*, for in *Strange Days* the signal from cortical recordings can be amplified to levels that cause fatal levels of excitation. Yet, the idea that television or a cortical recording could have a simple cause-and-effect relationship with the viewer is not to be taken literally. Instead, *Strange Days* and *Videodrome* can be used as a starting point with which to consider a range of interesting issues with regard to our relationships with mediated representations.

The visceral qualities of cortical recordings are emphasized in a scene where Lenny entices a new customer called Keith to sample cortical recordings. Whilst engaged in his sales patter Lenny tells Keith: 'This is not like TV only better, this is life – it's a piece of somebody's life – Its pure and uncut straight from the cerebral cortex. I mean you're there you're doing it, you're seeing it, hearing it, you're feeling it.' Keith asks Lenny what sorts of experiences he can offer via his illicit cortical recordings. Lenny replies that he can offer 'the things you can't have, forbidden fruit… I can make it happen, and you won't even tarnish your wedding ring'. So whilst using cortical recordings for personal purposes is illegal, Lenny gives the impression that they provide an important social function that enables people to taste the dark side of life, without resorting to actual violence, clandestine or perverted sexual relationships. Lenny continues, 'I'm your priest, I'm your shrink, I'm your main connection to the switchboard of the soul. I'm the magic man, Santa Claus of the subconscious.' From this we get the impression that Lenny is an entrepreneur who is providing a useful social function by meeting the needs of his client, rather than simply exploiting people's experiences for commercial gain. But Keith's fascination with playback also reveals the ways in which people may prefer immersive recordings of other people's experiences than the anxieties and messiness of actual social encounters.

Since the release of *Strange Days* the fascination with other people's lives appears to have continued unabated, particularly in relation to the popularity of reality television. However, information theorist Anthony Bryant (2010) points out that a common theme amongst reality television programming is that people are thrown together and encouraged to have intense yet disposable relationships.

In other words, these programmes often emphasize narcissism and self-gratification rather than care or concern for others. Furthermore, in recent years we have seen the exponential growth of user-generated media content. On this point, Nigel Thrift remarks that 'life-logging and digital curation are becoming important topics, as increasingly people record larger and larger amounts of their lives' (2005, 471). The important point here is that people are no longer just consuming the products of media companies; they are also making their own recordings or snapshots of their experiences for others to consume.

Yet user-generated content is not a straightforward recording of reality; it also involves an element of performance and impression management. For example, the producer(s) of such user-generated content may want to represent themselves in a favourable light to impress or they may want to provoke a response in others. The relationship between recording reality and performance is highlighted in *Strange Days* when Lenny approaches someone who is going to be involved in the production of a cortical recording for commercial use. Lenny tells the young man to build up the suspense during his performance because he does not 'want to be going out for popcorn'. What these comments indicate is that cortical recordings are not just raw extractions from the cerebral cortex; they also involve performance and draw upon cinematic and televisual codes, conventions and viewing practices.

The visceral and emotive aspects of cortical recordings are also relevant to the ways in which we interpret our experiences. The work of Antonio Damasio (2000) provides a useful way of interpreting cortical recordings in *Strange Days* since it addresses the relationships between consciousness and emotion. Specifically, Damasio seeks to explore how our emotions become known to us, in other words how they become consciously experienced. According to Damasio, emotions can be generated by direct conscious experience through our interaction with the world around us. In addition, emotions can be generated through a process of remembrance, imaging future scenarios or dreaming. In other words, various external and internal stimuli (such as instincts or unconscious processes) can trigger our emotions.

Expanding the debates on subjective experience, David Eagleman (2012) observes that we have a tendency to think that sensory experience involves processing information or stimuli from the world around us. Yet Eagleman contends that 'the brain is properly thought of as mostly a closed system that runs on its own internally generated activity' (44). Importantly, Eagleman states that some fine-grained tuning seems to take place as sensory data are modulated by subsequent neural processing. So there is a complex set of interactions between our sensory experience of the world around us and our internal embodied states. Discussing the implications arising from this argument, Eagleman makes the point that 'in this view, the difference between being awake and being asleep is merely that the data coming in from the eyes *anchors* the perception' (45). The ways in which we experience temporality is also a construction since by the time something enters our conscious awareness it has already changed, for our '...perceptual world always lags behind the real world' (51). Eagleman explains that subjective experience is similar to live television in which there is a few seconds delay between what happens and what is broadcast.

Lifeworld

In *Strange Days* cortical recordings can also be considered as a form of remembrance of something that has already taken place. Yet, it would be misleading to set up a simple division based upon the difference between subjective, immediate experience and technologically mediated experiences. Instead, Damasio and Eagleman's work indicates that our subjective, immediate experiences are intersected by a range of factors, including stimuli from the external environment and also internal stimuli such as memories or bio-chemical changes within the body.

Taking the points made by Damasio and Eagleman into consideration, we will explore the ways in which cortical recordings relate to external and internal stimuli in relation to a scene from *Strange Days* in which Keith samples a cortical recording of an 18-year-old girl taking a shower. In this scene, we see Keith gasping, moaning with pleasure and caressing his own body. In part, Keith is suspended from his real-time location, in a bar in Chinatown, to another moment, a fragment of someone else's life. But to what extent can Keith really experience what it is like for an 18-year-old girl to take a shower? Her experience of taking a shower is not an isolated one; it is a fragment of her life at a particular moment in time and in a specific social and cultural context. Furthermore, Keith's experience is technologically mediated since he must utilize a playback and headset to access it. Therefore, it seems to me that Keith does not really want to obtain in-depth insight into what it is like to be this young woman. Instead, he appears to combine his own fantasies about 18-year-old girls taking showers with the cortical recording itself. In this way, the girl on the recording becomes an object for his self-gratification.

Nigel Thrift's non-representational theory (2002, 2004, 2005), which highlights pre-cognitive, immediate and creative engagement with the world around us, raises some important questions about the possibility of capturing, storing and recreating a person's experiences using cortical recordings. Through undertaking a close reading of Thrift's work we can start to challenge the idea that Keith can somehow gain immediate sensory access to an 18-year-old girl taking a shower through playback technology. Although it may appear that the term 'non-representational theory' might be problematic when applied to a visual media such as film, Thrift's work provides a useful framework for the issues concerning representation, mediation and embodiment. In addition, Thrift does not oppose theoretical models which focus on the issues surrounding representation. Instead, non-representational theory embraces multiplicity, openness and creativity. Thrift adds that this approach is not about pinning down meaning; instead, it concerns 'the restless nature of the body's contact with the world, a contact which is not proximate but intertwined and continually coming to sense' (2004, 85). In other words, non-representational theory concerns spontaneity, emergence and the vitality of lived experience. From this perspective we can see that the issues surrounding lived experience are of particular relevance to the representation of cortical recordings in *Strange Days*. Specifically, Thrift's work challenges the idea that cortical recordings capture subjective experience at a particular moment in time and that this experience can be recreated, or re-experienced, by someone else at a different time and in a different setting.

What we find in Thrift's work is a relational model of reality which challenges ideas about the isolated, independent and purely rational subject who reaches out and shapes the world around them. Instead, his work emphasizes the ways in which experiences are formed through embodied interaction with the world around us. Phenomenology also offers a way of uncovering and exploring what Anne Buttimer and David Seamon term our 'lifeworld' which is 'the taken-for-granted pattern and context of everyday life, by which the person routinely conducts his or her day-to-day existence without having to make it constantly an object of conscious attention' (1980, 149). Making a similar point, Vivian Sobchack remarks that 'the perceiving and sensing body is always also a lived-body-immersed in and making social meaning as well as physical sense' (1994, 86). Taking these points into consideration, it is important to question the idea that cortical recordings could somehow capture and recreate a moment of someone's everyday life experiences and that this could be played back to someone else.

Our engagement in the world around us is shaped through our interactions with other people. On this point, Thrift remarks that we have the capacity 'to make models of the other' in order to ascertain the intentions behind their behaviour (2005, 466). At another level, interpersonal interaction involves socialization (of acceptable and unacceptable behaviour) and knowledge of cultural practices. We learn how to stand, walk and dress in certain ways and this may differ on the basis of socio-cultural attitudes towards gender or age. Elizabeth Grosz (1994) also reminds us of the gendered dimensions of embodied experience. As Grosz points out, there is a feminine-being-in-the-world that is also intertwined with specific social and historical contexts. Through socialization and cultural practices embodiment is shaped in terms of learning acceptable ways to sit, stand, walk and look at others. Griselda Pollock's (1988) trailblazing study of the ways in which femininity is defined in relation to the male gaze in *Vision and Difference* (1988) and John Berger's (1972) *Ways of Seeing* both point to the ways in which gender intersects the act of looking and who is looked at. The issues surrounding embodiment and spectatorship are particularly pertinent to Keith's enjoyment of watching an 18-year-old girl take a shower through playback technology. Drawing upon these theoretical ideas we can see that how a person moves, stands, talks and navigates social spaces is intersected by such factors as age, class, gender and race and the individual's life history. This means that a cortical recording is a fragment, a snapshot of the embodied experience of a specific person at a particular moment in time.

It is also important to explore the issues arising from experience in more depth because this is not as self-evident as it initially appears. Pragmatism is a useful framework for exploring experience through our practical engagement in the world around us. John McCarthy and Peter Wright state that 'for pragmatists such as [John] Dewey experience is more personal than behaviour; it involves an active self who not only engages in but actively shapes action' (2004, 54). However, 'in a world of change the same action can have different meaning and significance, as the context is always different' (2004, 71). Moreover, the person who is having an experience also changes over time, since as Thrift remarks 'the geography of each person consists of numerous layered subjectivities flowing through them' (2008, 85). Pragmatism is also helpful when considering the idea that cortical recordings capture visceral experience because this framework indicates that such experiences are context dependent. Since the recordings capture the particular context of one person but are played back by someone else, in a different context, this adds further doubt about the possibilities of re-living other people's experiences.

During their discussion of the cognitive turn in film theory, Richard Rushton and Gary Bettinson (2010) point out that when we watch a film we are utilizing the same sensory apparatus as we do in everyday life. Therefore, 'film viewing involves fairly automatic, non-conscious – or bottom up activities, such as perceiving figures in space and relating to physiological stimuli' (157). Yet, we also need to recognize that film 'engages us in cognitively more complex, top-down operations such as constructing plot events, storing information in memory, and forming empathetic attachments with characters' (157). *Strange Days*, like other films, removes objects, people and places from their spatial and temporal existence and re-orders them visually and temporally. The way in which the camera is positioned also stands in for the viewer's embodied movement through time and space. The viewer does not focus attention on the mechanism of the camera; instead, they notice the images that the film presents to them. This is especially noticeable in POV shots in which the realism of a film image is related to an individual spectator who is positioned in time and space. But this viewing position is abstracted from our phenomenological experience of the world around us. The film *Strange Days* comprises different images which are edited together to form a sequence, whereas we experience the world around us as a continuous temporal flow. We experience the world around us as a temporal flow as we move through space. As Vivian Sobchack (1994) points out, cinematic representation is about accumulation since a film unfolds sequentially and is a movement through time. Even when the film narrative involves flashbacks or flashforward sequences, it is experienced as an unfolding of meaning. On this basis we can interpret the representation of cortical recordings in *Strange Days* through our understanding of cinematic conventions, personal experiences and memories.

In his critique of technologically mediated experiences, Paul Virilio points out that 'real-time' technologies have nothing 'to do with our 'concrete presence' in the world instead they concern 'the elsewhere of a discreet telepresence' (2000, 11). Virilio adds that as such technologies become increasingly pervasive we are becoming enmeshed within telecommunications to such a degree that our bodies are becoming transmitters and receivers. According to Virilio, this intimate connection between embodiment and technologies is destroying our abilities to be fully present (in space) and to perceive the present (time). Virilio contends that what he terms the 'living present' is an experience that involves 'one's own body living in the here and now' (2000, 38). The tensions between technologically mediated experiences and co-present spatial and temporal relations are highlighted in *Strange Days* through Lenny's own use of playback technology.

Whilst Lenny offers his customers a taste of other people's experiences, he enjoys feasting on replays of his own prior experiences. He spends hours viewing cortical recordings of happier times with his former lover Faith (Juliette Lewis) and is unable to fully come to terms with the ending of their relationship. Lenny's compulsion for playback resonates with the discussion of Christian Metz's work on cinema by George Toles (2010). In particular, Toles recounts that for Metz '[n]othing is more deceptive than mere responsiveness to a pleasurable image' (160). Yet the pleasurable intensity of Lenny's use of cortical recordings to access his experiences with Faith are deceptive because they represent the past. Faith has found a new lover, but Lenny is still trapped in the past through repeatedly viewing playbacks of their former relationship. In one scene, Faith chastises Lenny, telling him that movies are preferable to playbacks because 'the music comes up, there are credits, and you always know when it's over'. Faith's comment highlights the ways in which cortical recordings distort 'real time' because they remove a person's experience from their original context and with this technology these experiences can be repeated over and over again.

Yet ethical issues such as care and concern for others start to emerge when Lenny watches an anonymous cortical recording that has been sent to him. The recording features Iris and shows her being blindfolded, raped and strangled to death. Iris's killer also uses playback so that she sees what is happening through his eyes. So, although Iris is prevented from seeing her killer face-to-face, she sees herself being raped and can feel his desire. Utilizing a series of cuts between Lenny's reaction to the recording, Iris's point of view and that of the killer also places the audience in an uncomfortable position, for we are invited to see this event from different perspectives: as a witness, the perpetrator and the victim. In his online review of the film, Roger Ebert (1995) states:

> It's revealing, how a scene like that seems so much more sad and distressing than the more graphic scenes of violence we see all the time in the movies. Bigelow is able to exploit the idea of what is happening; she forces her audience to deal with the screen reality, instead of allowing us to process it as 'routine' action.

After viewing the recording, Lenny is so repulsed that he vomits. This, I would argue, is a turning point in the film because after this Lenny seems to move away from cortical recordings and towards co-present forms of human connection.

After watching the cortical recording of Iris's rape and murder, Lenny seeks emotional support from his friend Mace (Angela Bassett), a tough, African American private security operative. When Lenny returns to Mace's home the scene cuts to a flashback and we see Lenny dressed in his police uniform reading a story to Mace's son Xander. This is the first time in the film that we see Lenny as a legitimate authority figure rather than a purveyor of illicit experiences. When the flashback ends Mace tenderly strokes Lenny's face. In this way the scene shifts our attention towards the fullness of embodied relationships in the same space and time. As Lenny forsakes his cortical recordings of former times with Faith we begin to see his loneliness, vulnerability and longing for authentic human connection.

Mace disdains the use of cortical recordings and regards Lenny's use of them as an addiction. It is only when Lenny tracks down the recording made by Iris on the night she was with Jeriko One and tells Mace that this has important social ramifications that she agrees to view it. As Mace puts on a headset, the scene cuts to a point-of-view perspective (based on Iris's point of view) which shows the shooting of Jeriko One. Although Lenny and Mace now have knowledge of what happened to Jeriko One and his entourage, viewing this cortical recording places them in a precarious position because they can identify the killers.

In the closing scenes of the film Lenny and Mace become separated in a crowd that has gathered downtown for a millennial party. Steckler and Engleman corner Mace, but since she is a trained security officer she deals with them swiftly and effectively, handcuffing them together. However, Mace's actions are observed by other police officers on the scene and what they see is 'a black girl crouching over two of LA's finest with a police baton' (Cameron, 1995, 180).[7] The police then start to brutally beat Mace and a crowd ominously gathers around her. By now the incident is on the verge of exploding into a full-scale riot. However, the situation is defused as Steckler and Engleman are exposed as killers.

The film draws to a climax with Lenny and Mace acknowledging their desire for one another, with an interracial kiss just as the clock ticks midnight. By this point Lenny's predilection for cortical recordings of Faith are forsaken for real-time pleasures with Mace. As Lenny and Mace kiss we also see the launching of balloons and festivities for the forthcoming millennium signifying a bright future. The killing of Jeriko One and his friends has been exposed and the principles of law enforcement are upheld. Even so, the idea that an interracial

[7] This scene also echoes the Rodney King incident mentioned earlier in this chapter.

kiss signals the start of greater understanding and tolerance within society seems rather ambitious. Furthermore, the ending of the film does not mark the end of playback technology; so whilst Lenny as an individual forgoes the lurid pleasures it offers, there are no guarantees that other members of society will follow suit.

This chapter has indicated that *Strange Days* evokes some of the social anxieties and unrest that took place in Los Angeles in the aftermath of the Rodney King incident in 1991. Indeed, from a wider perspective, the film draws attention to legitimacy and authenticity by provoking questions about the ways in which events such as the King incident are recorded and represented in the media. For to what extent do we witness such events through the media? And how do the media shape our understanding of events such as the King incident which have important social ramifications? In this chapter, I also contend that *Strange Days* alerts us to the problems that can arise when social division and inequality are combined with the lack of contact between different social groups, for what can arise out of this situation is antagonistic relationships that are rooted in fears of other members of society. This can create a vicious circle as the wealthier members of society start to consume privatized experiences of others rather than interacting with others who inhabit the same city.

It is also noteworthy that *Strange Days* is set in Los Angeles, a city associated with the American film industry and the production of popular cinematic fantasies and thrilling adventures. As discussed, *Strange Days* can also be regarded as a cinematic representation of a form of virtual reality, both in terms of the narrative theme and use of POV camerawork. In this way, the film is set in the context of the hype surrounding virtual reality technology during the 1990s, particularly in relation to the idea of experiencing compelling, immersive experiences. There are also intertextual resonances between *Strange Days* and other films that explored virtual realities and immersive experiences in the 1990s, such as *Total Recall* and *The Lawnmower Man*. In part, this discussion indicates that these films can be regarded as a way of dramatically exploring the implications arising from new technological developments, particularly in relation to subjectivity and memory. In addition, it also indicates that *Strange Days* can be interpreted as drawing our attention to the cinematic experience in other ways. Specifically, it shows that *Strange Days* concerns spectatorship, inviting us to see events from the position of the witness, the perpetrator and the victim. As argued, the use of POV camerawork in the film creates a compelling and emotive experience for the audience, since we not just are entertained but appear to be within the action that takes place.

Although *Strange Days* is set at a pivotal moment, on 31st December 1999, I have argued that many of its themes are relevant to contemporary debates concerning technologically mediated immersive experiences. Yet *Strange Days* does not present us with a technological deterministic view of the future of embodied social relationships. Instead, the film seems to suggest that it is how we use technology that is important, for as we have seen, cortical recordings can be used for different purposes. At present, it is still not possible to extract or playback our experiences using a technology such as cortical recordings. But the idea that our experiences could be captured and played back by someone else remains fascinating, though, as Thrift's work indicates, there is a vitality and spontaneity to our lived experience which cannot be simply recorded and recreated. Furthermore, drawing upon the work of Damasio and Eagleman this discussion also points out that our experiences are a complex mix of external stimuli and internal bio-chemical changes. From this perspective there is no such thing as an unmediated experience. What we see, interpret and understand involves our sensory perception, prior life experience and the material conditions in which they take place.

To sum up, this discussion has shown that *Strange Days* is a cautionary tale which remains relevant today. Indeed, since the release of the film, technologically mediated experiences have become more pervasive through the growth of social networking and mobile communication devices. Furthermore, the development of the device produced by Google which records our experiences in 'real-time' and enables us to share them with others through social networks can be seen through the prism of Virilio's conception of the human body as a transmitter and receiver. Admittedly, technological mediation allows us to reach out and touch others who are not physically co-present. Yet as the use of cortical recordings in *Strange Days* suggests, we need to think carefully about whether technological mediation of other people's experiences really does foster a sense of care and concern for others.

4

Virtual Identities and Material Bodies

This chapter critically analyses the relationships between online and offline identities and questions whether it is still relevant to make clear distinctions between them. When I first became interested in studying virtual environments in the mid-1990s computer-generated worlds were accessed through desktop computers and modems which were fixed at a particular location. Mostly this meant that if you wanted to access the Internet you were sat at a desk using a personal computer and a modem. Writing in the mid-1990s before the widespread use of the World Wide Web, tablet computers and smart phone technology, Timothy Leary (1994) proclaimed that as technology improves, a hierarchy of tactile experiences may emerge so that we reserve our embodied interactions for our most valued relationships, whilst the majority of our social interaction is technologically mediated in computer-generated environments. As the World Wide Web and other digital mobile technologies have become more widely available, it is interesting to return to Leary's point in order to consider how online interaction might impact our embodied relationships in more detail in the second decade of the twenty-first century.

At present, due to developments in telecommunication and computing it is possible to access multi-coloured, polyphonic, animated websites using mobile phone devices or lightweight portable tablet computers. Emerging from these changes are celebratory discourses which claim that these technological changes will allow us to easily connect to different people or explore different aspects of our identities and thereby improve our lives. But what is largely absent from such celebratory discourses are the problematic repercussions of technology, such as disconnection, and the anxieties that can arise when our actual physical embodiment does not measure up to our idealized virtual selves.

In this chapter, to explore the interconnection between online and offline identities in more detail, I will refer to a series of contemporary examples of virtual environments such as online games and social networks. In doing so, I will explore and interrogate contemporary films such as *Surrogates* (dir. Jonathan Mostow, 2009), *Catfish* (dir. Henry Joost, Ariel Schulman, 2010) and *Gamer* (dir. Mark Nevedine and Brian Taylor, 2009) since they dramatically represent some of the major issues surrounding embodiment, interactivity and engagement in online and offline environments – for arguably it is through the images and narratives that we find in the cultural realm that we attempt to make sense of social and technological change.

The chapter will also consider the argument that embodiment is not something that can be easily discarded or transcended through the construction of online identities. From this perspective, it can be contended that although we can extend our communicative reach, or construct different forms of identity online, our bodies still anchor us at a particular time and place (Hillis, 1999). Whilst navigating these different viewpoints, I will attempt to avoid overly simplistic arguments that regard embodiment as pure or authentic and online identities as inauthentic. Instead, I will insist that historical, social and cultural factors are important dimensions that intersect our embodied identities and relationship with digital media technologies.

To make such a wide topic manageable, the chapter focuses on three main theoretical approaches to online identities. The first approach is based on the idea that online identities offer fresh opportunities for developing aspects of our identities and new forms of expression and establishing relationships (Turkle, 1997; Carter, 2005; Guest, 2007). In this regard, online identities may be constructed through text-based communication, the development of a character in a narrative or computer-generated avatars. A vivid example of the construction of online identity and new forms of expression can be found in the MMORPG *World of Warcraft*. In *Warcraft* players can utilize elaborate avatars that represent races such as the Alliance or Horde; classes such as Druid, Hunter, Priest, Rogue or Shaman; and even different professions such as the blacksmith, tailor or alchemist. Through joining *World of Warcraft* or other similar multi-user computer environments, identity seems to become something which is incredibly malleable, though it could be contended that sites such as *World of Warcraft* are extensions of other forms of identity play such as Cosplay, where people produce elaborate costumes often in homage to their favourite film or television characters, or battle re-enactments. However, with digital technology changing our identity in sites such as *World of Warcraft* is much quicker and easier than sourcing materials and producing costumes for Cosplay or battle re-enactments. Through the use of digital imagery changing gender or body shape becomes something which can be done at the click of a mouse or choosing options on a drop-down menu.

The second approach to online identities is based on a series of arguments in which they are considered to be shallow in comparison to the richness of multi-sensory, embodied interaction with others (Wertheim, 1999; Franck, 2002). Indeed, it is important to be cautious about claims of unfettered online identities because they could result in what Lisa Nakamura (2002) terms 'identity tourism'. Nakamura's term 'identity tourism' suggests that the idea of playing or exploring different aspects of our identity through constructing virtual identities may be a temporary escape, or retreat from our everyday lives. In this regard identity tourism differs from the concrete social, cultural, economic factors that continually shape our everyday lives and the opportunities that are available to us. We may temporarily escape the embodied aspects of our identity through creating a Druid avatar on *World of Warcraft*, and may find this an enjoyable leisure time experience; but at some point we need to return to our physical body, which is located in a particular place and time. Furthermore, some people do not have Internet access or the material or symbolic resources that are necessary to construct online identities. There are digital divides because whilst the United States and the United Kingdom have a high density of Internet access and online usage, in other parts of the world access to the Internet may be limited due to technological, economic or political factors. And even in those countries that do have widespread Internet access, social distinctions can be made about the types of online networks we belong to, and how we use these sites to display our tastes and preferences (Papacharissi, 2009).

The third approach taken in this chapter will start to question the strict division between online and offline identities by placing changes in digitally mediated technologies in a wider social, cultural and economic context. In recent years new forms of mobile technologies have altered our relationships to mediated interaction, time and space. One of the consequences of the increasing portability of these devices is that we can access the World Wide Web as we simultaneously navigate the physical spaces around us (Moores, 2012). We may be in close physical proximity to other people but simultaneously engage with someone else via our mobile devices. So, one of the questions to be explored in this discussion is what happens when we can simultaneously construct and maintain our online identity alongside our physical bodies?

Whilst we may actively choose to construct and perform an online identity, we may at the same time be unaware of the digital footprints we are producing and how this can impact other parts of our lives. Therefore, it is important to explore the connections between online self-presentation, the flexible subject, and how this supports the demands of contemporary capitalism (Hearn, 2010). We also need to consider arguments relating to the ways in which the increasing use of social networking sites can be regarded as a way of generating massive and potentially lucrative data sets for marketing companies so that they can target advertising according to a person's online profile and how this relates to concerns about online privacy (Brooke, 2011; Andrews, 2013).

It is important to state at the outset that the increasingly complex navigation between online and offline identities is not something that is caused by technology alone. On the other hand, technology does make it seem easy to construct online identities and maintain online relationships. In addition, there are other factors to consider in relation to identity and social interaction, such as the increasing time pressure that many people face due to long working hours, commuting and the geographical separation of family members which make the ease of online communication seem attractive and convenient. But ultimately, the discussion will maintain that there is something valuable about embodied interaction and social relations that cannot be fully replicated online. Furthermore, arguments supporting the importance of embodiment are not to be dismissed as mere sentimentality. Although we may simultaneously navigate different aspects of our identity online and have different virtual selves, the discussion will reinforce the point that gender, race or class continue to be important factors that shape and can also impinge upon our everyday lives. In this regard the discussion points to the need for more extended debates around how we use technologies to construct online identities. Such debates may take place between individuals, through social and cultural organizations or in the political arena. Indeed, these debates will need to address important issues such as our ethical responsibilities to one another and the safeguards that are needed to ensure our safety and privacy when we engage in online activities. To begin the discussion, I will briefly mention some of the historical, social and cultural factors relating to online environments in order to set the context for the subsequent discussion of contemporary aspects of identity.

Online identities – virtual selves

One of the first online communities was established in the 1970s by the Computer Memory Project, based in Berkley, California. Lee Felsenstein, a founding member of the group, produced a computer terminal for public use in order to share information and ideas, such as where to find a guitarist for a band, or reviews of local restaurants. However, as Steven Levy remarks, there was also another dimension to the Computer Memory Project:

> By opening a hands-on computer facility to let people reach each other, a living metaphor could be created, a testament to the way computer technology could be used as guerrilla warfare for people against bureaucracies. (2001, 156)

Levy's statement indicates that there was a utopian aspect to the development of online communication. The idea was that through accessing online information, people would become socially and politically empowered. The Computer Memory Group was utilized via a public terminal, but by the 1980s individual access to the Internet could be achieved using a modem and a telephone line which were known as dial-up systems. For example, Usenet groups allowed people to read and post messages online (Rheingold, 1994). These messages were related to certain topics, thereby fostering a sense of an online community. Similarly, bulletin board systems such as the Whole Earth Lectronic Link (WELL) provided public message boards. Today, the WELL operates by membership and subscription services and provides access to information about conferences and discussion groups on a broad range of topics.

During the 1980s and 1990s the CompuServe service also enabled users to communicate online using text-based messages. It was during this period, as online identities were first coming to the fore, that journalist Lindsy Van Gelder wrote an influential article entitled 'The Strange Case of the Electronic Lover' (1991). At the time, CompuServe users could choose online names (known as handles) in a similar way to Citizen Band Radio enthusiasts, and Van Gelder discusses her interactions with an online regular called 'Quiet Lady'. As Quiet Lady's online identity developed, she began to disclose personal information to her newly found online friends. For instance, she revealed that her name was Joan Sue Greene and that 'she was a New York neuropsychologist in her early twenties, who had been severely disfigured in a car accident' (Van Gelder, 1996, 533). After the accident, Joan was left mute and confined to a wheelchair and withdrew from embodied forms of social interaction. However, she found solace through the many online friendships she formed.

As Joan's online friendships flourished, she changed her handle to 'Talkin Lady' to reflect her ebullient online identity. Van Gelder reports that over a two-year period Talkin Lady 'became a monumental on-line presence who served both as a support for other disabled women and as an inspiring stereotype-smasher to the able-bodied' (1996, 534). However, inconsistencies started to appear in Talkin Lady's online persona; for instance, whilst she had discussed the ways in which her facial disfigurements and physical impairment prevented her from engaging in face-to-face communication, she began to talk about her recent marriage. Talkin Lady even mentioned travelling abroad and going out socially with her husband, yet still failed to meet any of her online CompuServe friends.

Soon after it transpired that Joan was a man called Alex (though Gelder says that this was not his legal name), who was a New York psychiatrist in his early fifties. Subsequently, some of Joan's former online friends felt violated by the deception, since they had shared a great deal of intimate information with her. Van Gelder concludes that '[i]n the wake of the Alex/Joan scandal, some on-line habitués have soberly concluded that perhaps there's a thin line between getting out of one's skin and getting into a completely false identity – and that the medium may even encourage impersonation' (1996, 535).

Van Gelder's account of online identity highlights some key issues that relate to presence, absence and authenticity. In particular, the case shows that text-based online interaction can be used to construct and perform identity in sophisticated ways. Indeed, it appears that Talkin Lady had a lively textual online presence. Additionally, Alex's training and experience as a professional psychologist also meant that he was able to provide guidance to his online friends. Several of Talkin Lady's friends reported that 'she' had acted as a confidant and had often passed on valuable advice. On the other hand, the position from which this advice was given (as Joan/Takin Lady) was inauthentic. Furthermore, Talkin Lady's friends were unable to corroborate her online statements on the basis of her embodied presence and non-verbal cues. So the absence of the physical verification of identity facilitated the construction and maintenance of Talkin Lady's persona. After Talkin Lady was revealed to be Alex, 'her' online friendships disappeared, as many of her former friends reported that they felt used, even violated. Upon reflection it seems that Alex's construction of Talkin Lady was misleading but not necessarily malicious; rather, he sought to help other women, through online discussion. What Van Gelder's account highlights is the ways in which people were using text-based communication to construct and explore different aspects of identity. When Van Gelder's article was published, the construction of online identities was not yet a widespread phenomenon. However, the increasing availability of Internet access and usage has generated a variety of different ways of creating online identities through digital imagery and textual narratives which require further consideration.

A particularly perturbing aspect of the growth of online identities is the possibility of constructing fake or misleading online identities. As Andrew Keen observes, fake identities have 'become so widely adopted, they've been given their own term: "sock puppet", meaning the alter ego through which one speaks on an on-line community or posts on a blog' (2008, 76). One of the tensions arising from the increasing use of online identity and digitally mediated forms of communication is making distinctions about fake or misleading identities. So how do we evaluate the text-based or visual images that are used to construct online identities to ensure that they are genuine? How can we be sure that these identities are what they seem? These questions are particularly relevant in cases where we may never physically meet someone we are communicating with online on social networks or via multi-user role games.

On a disturbing note, Jaron Lanier (2011) discusses the darker side of online identities such as trolls who are abusive to others via social networking sites or blogs. In some cases the anonymity afforded by online identities is exploited in order to post abusive comments to others or even stalk them online. The anonymity of online communication may also mean that people are much more abusive to others than they would be if they were physically co-present, for it is much easier to post anonymous malicious comments without the threat of actual physical or verbal reprisal. The legal and ethical issues surrounding deceptive, malicious and unlawful forms of online identity are beyond the scope of the present discussion, but I would like to make a couple of preliminary points about them and signal towards important research in this area. For instance, there are concerns about the risks surrounding children's use of the Internet (Martellozzo 2012; Livingston et al. 2011; Livingstone and Haddon, 2009). Young children may be accessing social networks and could be misled by someone's online identity and lured into exploitative and unsavoury relationships with adults. Whilst there are anxieties around paedophilia in relation to children's safety in physical environments, the increasing use of social networks and other online sites raises a particular set of problems. For instance, there are problems arising from the proliferation of online imagery, the speed at which they can be circulated and how easily they can be accessed. There are, however, important differences between misleading online identities and those that are malicious, intentionally deceptive or unlawful. Admittedly, face-to-face relationships can also be built on deception, though embodied forms of deception may be dependent on many different factors such as the ability to lie convincingly, skilfully monitor non-verbal cues and cleverly utilize the art of disguise. Taking these factors into account, it seems that embodied forms of deception such as crossing gender or changing race are harder to construct and maintain than online identities. One of the ways in which an online identity can be constructed is through the use of avatars, which are graphical representations that stand in place of the embodied self. In the next section, I consider the use of avatars in more detail and the impact they may have on our online and offline identity.

Avatars

In the early days of the Internet online communities formed around role-playing games such as *Dungeons and Dragons* (Stone, 1995). Then, as the technology supporting Internet connections and computer graphics developed, computer-generated avatars (symbols) were used to represent online identity. As Kirsten Strayer (2010) points out, the term 'avatar' stems from the Sanskrit term *avatàra*, which means descent. In Hindu epics an avatar was considered to be an incarnation of a god who had descended to the earthly realm. In the context of contemporary culture, the term 'avatar' is commonly used to refer to computer-generated proxies for embodied interaction. For example, avatars are used in MMORPGs such as *World of Warcraft* and *EverQuest*.

Sherry Turkle was one of the first researchers to conduct a critical ethnographic study of avatars and online identities by visiting online multi-user domain games and other online sites of social interaction. In *Constructions and Reconstructions of Self in Virtual Reality: Playing in the MUDs* (1994) Turkle discusses a user called Peter, who played multi-user domains (MUDs) for 40 hours per week. According to Turkle, Peter constructed a virtual life that was more expansive than his offline, embodied existence. Online Peter was known as Achilles, which Turkle says was a projection of his ideal self. Achilles' virtual environment was elegant and fashionable; yet Peter's embodied situation was quite different, since he was a student living in a dormitory at the University of Massachusetts. In this way, Turkle's account of Peter exemplifies concerns about the disparities about online identities and the embodied aspects of our everyday life. But what happens when someone is spending a considerable amount of their time constructing and maintaining an idealized online self? For instance, how might such online pursuits impact upon a person's everyday physical interaction with others? And what happens when someone befriends Achilles and then meets Peter? To help answer these questions I will now turn to Denise Carter's ethnographic research into the virtual community known as Cybercity.

Carter spent three years studying Cybercity and interacting with its residents and found that 'relationships that begin online rarely stay there' (2005, 149). Instead, she found that many of the online residents met face-to-face and formed long-lasting friendships. What Carter found during the course of her research was that truth and consistency in online relationships develop over time, in a similar way as face-to-face friendships. Over a period of a couple of years, Carter, known online as Dutypigeon, formed an online friendship with a person called Zosma. Subsequently, Dutypigeon and Zosma decided to meet in person. Since they used avatars in Cybercity, Carter and Zosma decided to exchange photographs before they met, so they knew what one another looked like. In this way, the photographic image was used as a means of verifying identity, a practice that is familiar through the use of passport photographs. What is interesting, in terms of this discussion, is that Carter states that she was thankful that Zosma had been truthful and that 'she was exactly like her photograph' because 'it was not until I saw her that I realized how important that particular truth was' (163). Carter adds that 'in that instant of meeting face-to-face I realized that we had maintained an authentic perception of each other, and that failure to do so would have destroyed our friendship' (163). Soon after their face-to-face meeting Dutypigeon and Zosma decided to meet up with another online friend Phat. However, they found that Phat:

> Was about 10 years older, and a stone heavier than the photograph we had exchanged. Talking later, Zosma and I agreed he was a bit vain to send us a younger photograph, but also that he was exactly the same person we knew. Therefore, in this case we maintained our own perceptions of him even though he had not been completely truthful. (163)

What I find interesting is that the trust that Carter and Phat had developed over time mitigated his use of an old photograph. This chimes with Carter's main argument that developing and maintaining friendships is a multifaceted process involving trust and disclosure, often over a prolonged period of time. In other words, because Phat was a trusted friend, Carter overlooked the ways in which he used a misleading photograph to represent himself.

Another example of the use of avatars in online environments is *Second Life*, which was developed by Linden Labs and launched in 2003. *Second Life* offers the opportunity to construct identities through the use of avatars and interact with others. It even has a form of economic exchange, known as Linden Dollars, which can be exchanged for actual currencies such as the US dollar. Some people opt for a standard avatar in *Second Life* (from a limited range) that includes people, animals, robots and vehicles, but others choose to purchase additional 'virtual' items, such as clothing, shoes, hats or glasses to customize the appearance of their avatar. In some cases, *Second Life* residents purchase their own custom-designed habitats such as a house or an apartment. A *Second Life* resident may find it relatively easy and enjoyable to construct their online appearance and lifestyle through the use of digitally malleable imagery. At the same time, other *Second Life* residents may benefit financially from offering their digital design and construction services to others within this online world. Yet there are also differences between social and economic exchanges within *Second Life* and our current global economic system. For instance, after the global economic crisis of 2008 many US citizens lost their jobs and found themselves in crushing levels of debt which was often the result of taking out risky mortgage products in a financially inflated housing market. In this instance, there are important differences between the ease and attractiveness of constructing elaborate online identities and environments on *Second Life* and the actual foreclosure of homes across the United States and in other parts of the world, such as Spain and Ireland.

Tim Guest's (2007) account of communicating with a *Second Life* avatar known as Wilde Cunningham also offers further insight into the ramifications of online identity and embodiment. Specifically, Guest outlines the ways in which Wilde Cunningham was constructed by nine residents of the Evergreen Center, near Boston. After communicating for some time in *Second Life*, Guest decided to meet the residents of the Evergreen Center in person. Upon meeting these residents he discovered that many of them had cerebral palsy or had other special needs and had constructed Wilde Cunningham to explore other facets of identity. According to Guest:

> Wilde's stop through the electronic looking glass had turned a negative into a positive. For the nine, mired in their respective physical limitations, leaving the world behind – if only for a while – was the most exciting thing in the world. (2007, 54)

However, the residents were supported by their care worker who transposed their verbal utterances and non-verbal cues into online communication via the avatar Wilde Cunningham. In this way their care worker not only facilitated but also shaped their online interaction. It is also noteworthy that their care worker had the cultural capital required to construct and maintain Wilde Cunningham's identity, from her previous experience of playing computer games such as *The Sims*. Having resources such as a dedicated computer room, as well as the finances required to acquire computers and Internet access, also contributed to the residents' use of *Second Life*, and this may not necessarily be available to other residents in care homes or assisted living facilities.

Although Turkle (1994), Carter (2005) and Guest (2007) present some of the positive uses of online interaction, identity construction and performance, other writers present a more critical stance. For example, writing around the turn of the millennium, Karen Franck (2002) and Margaret Wertheim (1999) made clear distinctions between virtual and embodied identities. For instance, Margaret Wertheim states that 'to play at being a singing fish or the opposite sex can indeed be a positive experience, but to believe that these experiences are equal to life in the flesh is a delusion' (1999, 246). Presenting another dimension to this issue, Lisa Nakamura (2002) cogently argues that in some cases the naming and appearance of avatars perpetuate existing racial and ethnic stereotypes and upholds the idea of the exotic or otherness. Furthermore, as Nakamura points out, people may use avatars to change their race, but this does not have the same consequences as embodied social interaction which is intersected by visible and non-visible aspects of difference. Despite the strides that have been made in recent years to promote equality and diversity and the legislative changes that have taken place, especially in the European Union, gender, disability, race, class and sexual orientation continue to be important factors that both shape and limit our embodied social relationships and the opportunities that are available to us.

One of the appealing aspects to online identities is that they can be a way of constructing an ideal self. For example, through the use of avatars, it is possible to shape our online appearance in all sorts of ways, through size, shape, clothing or accessories. In contrast, changing our physical bodies through diet or exercise takes a great deal more time and effort. For some the process of constructing an online self can be a way of projecting an ideal self to attract or impress the other people we encounter online. But what happens when there is this gap between our ideal online identities and our actual physical appearance? Furthermore, if a person spends most of their time engaging with their online selves how might this impact their physical bodies and the social relationships they have with their families, friends and co-workers? To explore these complex issues, we will now turn our attention to the film *Surrogates*, since this draws attention to some of the implications arising from the use of ideal, technologically constructed bodies.

Surrogates

The main premise of *Surrogates* is that human beings have produced idealized robotic avatars that act as substitutes for human embodied interaction. Whilst these robotic avatars are used in professional and personal situations, humans are rarely seen in public. The ways in which surrogates are regarded as superior to humans is illustrated by the use of the term 'meatbag' to refer to the human body. The film also depicts different types of surrogates; there are specialized surrogates such as those that have been built for law enforcement agencies and budget models which are not able to display complex facial emotions. Consequently, some people can purchase advanced models which act as a mark of social distinction whilst others are judged negatively because they can only afford a basic model which limits their capacity for effective social interaction.

Surrogates are also used in extreme sports and in combat situations because if a surrogate is harmed its human operator will remain unscathed. But, whilst surrogates are sophisticated machines, they lack self-consciousness and the responsibility for their actions remains with the human operator who is controlling them via a massive online system. Even so, the use of these robotic substitutes means that humans acquiesce to the extensive technological mediation of daily life. Put simply, humans can no longer engage effectively in social situations without having a surrogate.

The opening sequence to the film includes pseudo news-style footage detailing the development of surrogates. What is noteworthy is that this footage shows how representational conventions within the media, such as objective forms of news reportage, can influence our view of truth and reality. The footage is also punctuated by rapidly edited infomercials for Virtual Self Industries, the company that produces surrogates, a point which raises issues such as the impartiality of news reporting and the relationships between television networks and commercial interests. Included in the news sequence is war footage showing the use of cyborg soldiers who are equipped with biometric facial recognition and increased body armour, thereby pointing to the military and industrial connections to such technologies, which is something Donna Haraway (1991) has repeatedly emphasized in her work on cyborgs. Nested within this pseudo-news footage is a clip of the actual research conducted by Professor Hiroshi Ishiguro from Osaka University. In 2008, Professor Ishiguro developed a robotic copy of himself known as Geminoid, and more recently he developed a female robot known as Geminoid F, which is able to realistically mimic human facial emotions. In this regard, it seems that the inclusion of Professor Ishiguro's work in the film's opening sequence creates the impression that surrogate technology will be available very soon.

The film also details the ways in which a group named the Human Coalition, led by Zaire Powell (Ving Rhames), known to his followers as the Prophet, oppose the use of surrogates. At the start of the film the Prophet issues a warning against the use of surrogates, stating:

> Look at yourselves. Unplug from your chairs, get up and look in a mirror. What you see is how God made you. We're not meant to experience the world through a machine. Those machines walking around out there are a lie.

The Prophet's statement suggests that the Human Coalition's resistance to surrogates is underpinned by religious notions of God, rather than humans, as a creative force. According to this line of thinking, humans must not usurp God by creating synthetic beings, though it is unclear whether other members of society or even those within the Human Coalition might object to the use of surrogates on other grounds apart from religion. There is a short clip in the film which depicts the coalition rallying against surrogate technology and successfully petitioning for the introduction of surrogate-free zones in some US cities. Yet the Human Coalition is represented as a minority movement since the majority of the US population uses surrogates.

Yet there are also elements in the film which undercut the widespread use of surrogates and cast doubts upon replacing human embodied interaction with robotic substitutes. The creator of surrogates, Dr Lionel Canter (James Cromwell), for example, comes to renounce their use and hacks into the system that connects humans and surrogates. This results in an unprecedented situation whereby the harming of a surrogate leads to the death of its human operator. In this way, by changing the parameters of the system, Canter's intervention highlights the issue of human responsibility and mortality.

At the end of the film the protagonist Tom Greer (Bruce Willis) temporarily halts the computer system that links surrogates and their operators. By doing so he gives humans the opportunity to return to their corporeal existence. However, the film does not imply that the interruption to the system will necessarily lead humans back to corporeality on a permanent basis. Furthermore, it transpires that the Prophet, unbeknown to his followers, is a surrogate and that Canter is his human operator. In this way the deceptive use of a surrogate as the leader of the Human Coalition poses serious doubts to the ethos of the resistance movement against surrogates.

In many ways, *Surrogates* extrapolates from the existing use of avatars in online environments to present a futuristic scenario. In this regard, *Surrogates* can be interpreted as a cautionary tale of what could happen if technologically mediated social interaction becomes the norm and social interaction based on embodied presence is diminished. After considering the use of idealized selves in the form of surrogates, we will now consider the ways in which the ability to access the Internet and construct online identities via social networks is intersected by economic concerns, social positioning and performance.

Social networks

One of the most well-known social networking sites is Facebook, which was established by Mark Zuckerberg whilst he was a student at Harvard University. Zuckerberg's initial attempts at developing a social network included Course Match, which allowed students to view the courses other students were taking, and Face Mash, which offered students on the Harvard campus the opportunity to rate the attractiveness of others. When Zuckerberg launched the Facebook (as it was originally named) in 2004, it was an exclusive social network for Harvard students only, but this soon extended to other campuses in the United States. Now Facebook has extended to other countries around the world. One of the reasons why Facebook is attractive to potential users is that it is possible to set up a free profile. After setting up a profile it is also possible to upload photographs, use instant messaging and link with others to establish online friendships.

Computer scientist Jaron Lanier (2011) is critical of social networks, arguing that the use of pre-existing, drop-down menu choices (such as relationship status) and a structured profile format results in reductive representations of identity. On the other hand, Zizi Papacharissi (2009) contends that the format and structure of social networking sites can be customized by users and are not quite as limiting as Lanier suggests. For example, people may use different photographic images or links that display their interests and these provide some scope for individual expression. Some young people are hyper-vigilant about maintaining their online identity and overly concerned about how their Facebook profile is viewed by their peers. Consequently, they may attempt to represent themselves in a flattering light through the use of professional or digitally enhanced photographs. They may also utilize carefully constructed statements about their background, interests and activities. In some cases, a person may agree to be friends with someone they already know through other circumstances (either at school, university or work), but in other cases they may have friends they have never actually met. So to what extent are these online connections with someone you have never met before a form of friendship? And what are the drawbacks of such online relationships? For instance, how can we be sure that the person we have befriended has a genuine profile on Facebook?

Another major social networking platform is Twitter, which was founded circa 2006. Twitter enables people to set up a free account, profile and post and respond to messages known as tweets that can be a maximum of 140 characters. One of the advantages of Twitter is that it is easy to compose and read these succinct messages using mobile technologies. The ease with which Twitter users can tweet has resulted in some people publishing the minute details of their actions, experiences and thoughts, often blurring the distinction between the public and the private.

Meanwhile, LinkedIn is a professional, career-orientated social network that enables users to create their profile in a CV-style format that emphasizes their qualifications and previous work experience. From this perspective LinkedIn can be regarded as an extension of existing business practices such as referrals, recommendations and networking events. In contrast to Facebook, Twitter and LinkedIn, ASmallWorld is an exclusive social network that operates by invitation only and consists of wealthy and influential members. Of relevance here is Papacharissi's remarks that 'the mere act of joining a particular network represents a taste statement' (2009, 212). Furthermore, 'membership of an online professional network communicates a statement of class and profession' (212). Papacharissi's comments cast doubts about claims that online identities are a way of transcending the social and cultural markers that intersect our embodied existence. Instead, social networks can be regarded as a way of making social distinctions based on the types of networks we belong to, and how we use these sites to display our tastes and preferences.

In recent times, within hermeneutic fields of enquiry, there has been an emphasis on identity as an ongoing process rather than an inherent quality.

These ideas have found expression in arguments about the construction of subjectivity and the de-centred or nomadic self (Foucault, 1994; Braidotti, 2002; 2006). Whilst our body seems to anchor us to a single identity, it is also recognized that we engage in different performances of the self in accordance with social roles and cultural conventions (Goffman, 1959). In the past we have performed those roles in shared physical spaces such as the park, office, social club, gym, café or bar. In these environmental contexts, relationships were built on the basis of interpreting and responding to non-verbal cues, facial expression, hand gestures, tone of voice, levels of eye contact and embodied proximity. However, American communication scholar Nancy Baym (2010) contends that it is also possible to engage in online self-presentation through customization by using emoticons, capital letters and informal language. Consequently, Baym contends that 'socioemotional communication may be easier face-to-face, but it is common and successful in digital media as well' (2010, 103). Yet there can also be important temporal differences between online and embodied forms of communication. In the case of synchronous face-to-face communications, there is a shared sense of space and immediate responses and diversions can also take place, whereas online asynchronous forms of communication such as writing a blog or responding to someone else's blogs or postings may allow us to carefully choose how we respond. A further point is that unlike symbolic emoticons, emotional facial expressions are inextricably linked to physiological changes in the body. Therefore, when we are conversing face-to-face with another person we are not just responding to the content of their words; we also pick up on other cues such as the emotional quality of their facial expression and tone of voice.

On a critical note, Alison Hearn (2010) makes connections between online self-presentation and contemporary capitalist modes of production. In particular, she focuses on the type of subjectivity this generates. Hearn points out that the rapid and global flow of contemporary capitalist production generates instability in the labour market as multi-national corporations can now move their production and distribution operations to those places that offer the most favourable economic conditions. According to Hearn, these conditions generate the idea of the flexible personality, which is used to describe someone who can quickly adapt to changing circumstances. Consequently, 'we see a shift from a working self, to the self as work in the form of a self-brand with reputation as its currency' (2010, 426). Furthermore, 'the self as commodity for sale on the labour market must also generate its own rhetorically persuasive packaging, its own promotional skin, within the confines of the dominant corporate imaginary' (427). Arguably, the idea of self-presentation is also linked to the ways in which social positioning on the basis of class or profession is becoming increasingly intersected by other markers such as leisure and consumption. One way of producing a favourable form of self-presentation and self-promotion is via social networking sites, by carefully constructed images and narratives about our leisure time interests, activities and patterns of consumption.

Whilst using social networks may be something that we associate with our leisure time activities, such as contacting friends, constructing and maintaining our online identities can be regarded as a form of labour. On this point, Hearn remarks:

> just as women's work in the home contributes directly to capitalist accumulation but is also 'not counted' in formal economic analysis, the new on-line economy is entirely under-girded by free labour, even as dominant discourses work to cloak its existence under the rubric of 'authentic expression' and 'personal empowerment'. (434)

Hearn's statement illuminates the ways in which the labour that is involved in self-presentation in social networks is enmeshed within contemporary capitalism. For example, displaying our tastes and cultural knowledge through our online profile provides valuable marketing data for commercial organizations. In this way there are complex and in some cases troubling connections between the online virtual selves we create and the impact this can have on other areas of our life as we are targeted by sophisticated advertising that is tailored according to the tastes we express on social networking sites.

A rather different approach is presented by media and communication theorist David Gauntlett, who dismisses critical views towards Web 2.0, contending that some of them are 'relatively predictable' and are made by those with 'rather inflexible views' (2011, 185). However, since he does not elaborate on these remarks we are left wondering what these views are, what makes them predictable and who has such inflexible views towards Web 2.0. Furthermore is a refusal to wholeheartedly champion Web 2.0 simply due to an inflexible mind-set or could this exemplify valid and genuine concerns? Gauntlett moves on to consider arguments relating to the ways in which user-generated content can be a source of important and profitable information for corporations such as Google, YouTube and Facebook. According to Gauntlett, 'In this view, the site owners are the lazy greedy factory owners, whose wealth is built upon the hard labour of thousands of creative workers' (187). Gauntlett has a valid point when he says that it can be misleading to apply a Marxist framework of the forces and relations of production to user-generated content on social networking sites. Indeed, Gauntlett goes on to challenge this position, arguing that there are important differences between the labour of someone who works in a factory churning out mass-produced goods and the creative activities that are part of user-generated content on social networks. Gauntlett contends that those who use social networks as a creative platform 'never hoped or expected to be able to make any financial gain from sharing their work' (187). In addition, whilst people can make something and share it via social networks, this is not the same as making physical, mass-produced commodities. For the creative activities that are shared via social networks may not involve the transfer of material objects and ownership in the same way as industrially produced goods.

Gauntlett concedes that the labour that is involved in creative production on social networking sites is 'harvested' by the companies that run these online platforms but that it is 'happily and voluntarily given by users who want to share their creative work' (188). Consequently, he states that 'the creative social acts which we see the results on YouTube are not the equivalent of work, which you would reasonably expect to get paid for' (189). Gauntlett has a point when he says that we do not usually think about putting together a photo album in the same way as paid work. However, some people may engage in creative pursuits on social networking as a way of presenting themselves in a favourable light to potential employers. Therefore, demonstrating their creative skills online, for some people, is a vehicle towards employment or other form of remuneration. Furthermore, depending on our privacy settings, if we post our photo album online via social networking sites then information about who views those images, or how they are tagged (categorized), can be valuable in terms of marketing and advertising companies.

Although Gauntlett acknowledges some aspects of the arguments pertaining to data mining and surveillance online, he seems to sweep them aside. In fact, Gauntlett says these arguments are 'undermined by the black and white worldview of the academics who write about these matters, who wear their single-mindedly "critical" stance as a badge of honour' (193). But critical thinking is not about holding a black-and-white worldview; instead, it is about considering issues from a range of different perspectives. On a rather defensive note, Gauntlett writes, 'people who say descriptive or positive things about aspects of web 2.0 are dismissed as naïve optimists' (193). Although Gauntlett raises a valid point when he says that in some cases the positive aspects of Web 2.0 are underplayed, critical analysis provides a useful way of exploring a range of advantages and disadvantages to Web 2.0.

Gauntlett defends his argument about the creative potential of social networking by stating that if people 'felt that they were being punished or exploited they would simply do something else' (188). But to what extent could they use online platforms that have the same social recognition and market penetration that is offered by Google, YouTube or Facebook? Some of those who use social networks to construct online virtual selves may find such activities incredibly desirable and enjoyable. Yet at the same time, they may be largely unaware of the consequences arising from their online profiles and the extent to which this becomes a valuable commodity in the capitalist economy. So whilst joining a social network and constructing an online profile appears to be a free service, those who sign up to such platforms are required to adhere to the terms and conditions of the commercial companies who provide them. From this perspective those companies have considerable power in terms of providing access to their service and what they do with the information that is provided by people who create their online profiles.

In sum, claims about online identities empowering the individual are also undermined by some of the processes involved in self-presentation. For although someone can customize the ways in which they are represented online, they may also be disclosing information that is of commercial value. Furthermore, it is important to reiterate that users need certain material resources such as Internet access, a computer or mobile device to construct their online identity. In addition, a certain amount of skills or what Pierre Bourdieu (1986) calls cultural capital may be required in order to construct and maintain online identities. Of note here is that these material resources and cultural capital are not equally available to all members of society.

I now want to bring us back to the point that online identity is a technologically mediated experience. In some cases we know what our friends look like and how they behave when they are physically present and we compare and contrast this with their online profile on Facebook, or Twitter. In other cases we may engage with someone at the level of appearances online and in due course other aspects of their identity may be exposed or uncovered (Gunkel, 2010). The following critical analysis of the films *Catfish* and *Gamer* elaborates some of the issues arising from the differences between online profiles and other aspects of identity in greater detail.

Catfish

The film *Catfish* introduces viewers to Yaniv Schulman, a twenty-four-year-old photographer from New York who receives a painting from Abby, an 8-year-old artistic prodigy from Michigan. After receipt of the painting, Yaniv becomes entangled in a web of online relationships with Abby, her sister Megan and her mother Angela. Over time, Yaniv speaks to Megan on the telephone and they also begin to send regular text messages to each other. Gradually Yaniv and Megan become romantically involved. As their relationship intensifies, Yaniv decides to turn up unexpectedly at Megan's home to meet her in person. However, it transpires that Megan is a false identity which has been constructed by Abby's mother, Angela Wesselman-Pierce, who is a middle-aged woman who cares for her disabled stepson. In addition, Angela is the artist who produced the paintings which were allegedly produced by her daughter Abby.

When constructing Megan's Facebook profile, Angela utilized an image of a photographer and model named Aimee Gonzales, which was easily obtained online. Overall, the film seems to represent Angela in a sensitive manner and gives the impression that her construction of Megan's identity was a way to alleviate some of the troubling aspects of her daily life. In this way the film suggests that Angela's online activities were not malicious, rather she was seeking excitement, attention and even romance online. Furthermore, it appears that Angela was not attempting to use Megan's fake identity for monetary gain or notoriety. On the other hand, as Roger Ebert (2010) recounts, there are questions about the authenticity of the *Catfish* film. For it is possible that Yaniv was aware from the outset that Megan's identity was inauthentic but proceeded to enter into a relationship with her family for the purposes of making a film. Also, since Yaniv is a professional photographer to a certain degree the film generates publicity for him. It could even be argued that Aimee Gonzales has benefited from the publicity generated by the film because the web page which promotes her activities as a photographer and model also includes links and commentary about the *Catfish* film.

Catfish highlights the ways in which a false Facebook identity can be unravelled because it is not anchored in reality. Yaniv starts to suspect that Megan is not what she seems after attempting to verify some of the information she has provided online. What happens is that over time a series of gaps starts to emerge between Megan's online profile and reality. Ultimately, the discrepancy between Megan's online identity and reality is uncovered when Yaniv attempts to meet her in person. *Catfish* also indicates that the unravelling of an online identity can be a very emotional experience, for when Angela confesses that she constructed Megan's profile she breaks down in tears. Taking these points into consideration we can see that *Catfish* can tell us a great deal about the continuing importance of embodied personal interaction and trust.

Gamer

Gamer is a futuristic film set in the year 2034 at a time when the human brain can be interfaced with computer technology, thereby disrupting the boundaries between online and offline environments. Arising from these technological changes is a popular network called *Society*, which enables human beings to become puppets that can be played by other users. Those who allow themselves to be played do so for financial gain so that human embodiment becomes a commodity to be bought and sold. Furthermore, there is no dialogue between those who are played and their players. In this way, those who are played appear to be lacking agency since they are represented as mute avatars that can be manipulated at will by their players. For instance, Angie (Amber Valleta), a young, blonde, attractive woman, allows her body to be played by Gorge (Ramsey Moore), an overweight man who seeks sexual thrills. Through the use of Angie's body in *Society*, Gorge experiences what it is like to be conventionally attractive and sexually desirable. In this way, *Society* starts to blur the distinctions between online and offline environments since Angie is physically embodied but her mind and actions are controlled across a technological network by Gorge.

After the phenomenal success of *Society*, its creator Ken Castle (Michael C Hall) produces a new game called *Slayers*. In this game death row convicts are offered the opportunity to play for the ultimate prize: their freedom. *Slayers* is played in real time, with live action footage which is broadcast to a huge audience. Like *Society*, the convicts in *Slayers* become human puppets who are controlled by other players, but their actions have serious consequences, for they use real weapons and can be killed. As the film progresses it becomes evident that the parameters of *Slayers* are skewed so that it is unlikely that any of the convicts will achieve their freedom. A renegade group called Humanz, however, instigates a virus into *Slayers* so that its most successful player Kable (Gerard Butler) can communicate with his user, a young gamer called Simon (Logal Lerman). Kable then persuades Simon to forgo his agency as a player which ultimately results in Kable's escape from *Slayers*. However, at the end of the film it is unclear if *Slayers* will end and another death row convict will be recruited to replace Kable. Of note here is that when players are killed in the game this increases the ratings for *Slayers* and in this way the film provides critical commentary on emotionally driven reality television and the commercial importance of ratings to advertisers.

To sum up, *Gamer* raises important questions about performance and the possibility of making distinctions between online and offline environments. *Society* presents the radical commodification of the human body as something that can be bought and sold for financial gain and that the human body itself can be an avatar for someone else. In addition, there are grave consequences to *Slayers* since performing in this game could result in death. At the same time the loss of a human life in *Slayers* becomes a commodity because it raises the emotive stakes and draws in audiences.

At present it is possible to conduct some aspects of life using digital technologies such as online banking, playing MMORPGs or social networking. But these activities are fundamentally underpinned by the sensory motor features of the human body. It is the sensory motor capacities of the human body that enable and constrain our adaptation and orientation in the world (O'Reagan, and Noe, 2001). Indeed, the sensory motor features of the human body such as sight, sound and touch undergird online interaction. For example, we are able to look at other people's profiles on Facebook or listen to their music on My Space, and we use the sense of touch to type messages, or to click a mouse.

Mediated lives

Although we can explore different identities online, we still need to attend to our physical offline needs. But in some extreme cases people have neglected their physical needs when they have become highly absorbed in online games. For example, Guest (2007) investigated online addiction in South Korea and discusses the case of Lee Sung Seop, who had a cardiac arrest and died after playing *World of Warcraft* for 50 hours virtually non-stop. Guest adds that in 2005, '10 South Koreans died in similar ways' (2007, 308). A disturbing report by Mark Tran (2010) also outlines the ways in which a South Korean couple (known only as 'the Kims') became so absorbed in Prius online (an MMORPG site) where they tended to a virtual young girl that they neglected their own three-month-old baby. Due to their excessive attention to Prius, the Kims' real baby suffered from malnutrition and died of starvation. These examples point to the serious consequences that arise from prioritizing online identities and disengaging from our physical embodiment.

As discussed in the 1980s and early 1990s, Turkle was favourable towards the new opportunities offered by computer technologies to explore different aspects of identity. But by the late 1990s she was already starting to have doubts about her initial optimism. As Turkle cogently argues in *Alone Together* (2011), it is possible to spend a considerable amount of time controlling our online performances, such as uploading flattering photographs of ourselves, or tweeting witty remarks. But real-time face-to-face communication is rather different because there is an element of the unknown. Face-to-face conversation can be thought of as a form of improvisation whereby we may riff from different points in response to what is said. Since when we are having a face-to face conversation we are never quite sure what the other person's response will be, or how the thread of the conversation will unfold, or change direction. One of the disconcerting findings of Turkle's study of technologically mediated forms of communication was that some young people actually fear the unpredictability of face-to-face conversations. Instead, in some cases young people prefer to engage in mediated forms of communication with their peers because they can construct what they want to say and edit it before their message is sent.

In agreement with Turkle's argument, if we spend a lot of time controlling our online relationships and managing our self-performance, it will take us away from other things. For instance, our online relationships may be taking us away from connecting in a meaningful way with those who are closest to us (spatially and emotionally), such as our family members. Initially, online relationships appear attractive; they hold out the promise of overcoming loneliness and isolation by connecting to other people. However, problems occur when we start to prefer online relationships to face-to-face relationships in the world around us, since we may be deluded or misled by carefully constructed modes of communication and idealized self-presentation. For, we may be deluding ourselves through constructing an idealized virtual online identity in order to impress others and we may also be misled by their online identity.

Digital media allow us to 'go' places, to be present at a distance, and in this way physical environments and media environments overlap. According to Shaun Moores, 'telephone use, and electronic media use more broadly, is best seen as "pluralising" social settings, as opposed to removing people from one type of situation, which becomes marginal, and putting them in another' (2012, 6). But what impact does this pluralizing have? Is it really possible to navigate effectively between different physical and media-related environments on a regular basis? In recent years we have seen that public places such as trains or bus stations, parks or cafe are no longer places where we encounter others and connect with them. Now, these are places where we may be physically present but also disconnected from one another through listening to an iPod, using mobile telephones or tablet computers. But when we use technology in this way we split our attention, simultaneously navigating online and offline identities and relationships. What I find most problematic is that when we engage in this type of behaviour, not only are we not fully physically present at a particular time and place, but we can also treat others as if they are not really there. And if repeated often enough this type of behaviour may significantly erode our ability to connect and empathize with others who are physically co-present.

A vignette from Turkle's (2011) study about Pete, a middle-aged, married man with two small children aged four and six, vividly illustrates the drawbacks of simultaneously engaging in online and offline identities. On *Second Life*, Pete has an avatar called Rolo who is young, attractive and has 'married' an avatar called Jade. Yet Rolo and Jade have never met in real life. According to Pete, he is able to confide in Jade but finds it difficult to talk to his real wife Alison. Whilst Pete's children are having fun in the playground, Pete uses his mobile telephone to be online with Jade. Turkle's point is that we may be living out different aspects of our lives simultaneously, and this is what is new about present-day technology. Pete is not in a hotel room with his lover whilst his children are at home with their mother. Instead, Pete is physically present with his children but his attention is elsewhere. When Turkle questions Pete about his behaviour, he refers to his 'life mix', explaining that this 'is the mash up of what you have on and off line' (2011, 160). The vignette about Pete illustrates the ways in which people can now engage in multiple and fragmented forms of self-presentation and interaction. But what will be the long-term impact of Pete's behaviour, with respect to his children and his marriage to Alison? For, it seems that Pete is showing his children that engaging with his online relationship with Jade is more important to him than watching them. Consequently, Pete is missing out on precious experiences with his children. Furthermore, he is attempting to be in two places at once, splitting his attention between real-time physical relationships and engaging with the flickering signifiers of Jade's virtual online identity.

In contrast to Turkle's warnings about the loss of real-time conversation and the richness of physical contact, Baym (2010) discusses the ways in which threats relating to different technologies have been made in the past. For example, she discusses the ways in which Socrates thought that the alphabet and the use of written language was a threat to oral culture and mnemonics. Since the time of Socrates, the flourishing oral culture of Ancient Greece has declined, yet new cognitive skills have come to the fore. So Baym's point indicates that cognitive abilities and social interaction are shaped by particular cultural and technological factors and they also change over time. As discussed, in technologically developed cultures there are now myriad forms of technologically mediated communication, including email, instant messaging, texting and blogging. It seems to me that the sheer volume and ubiquity of these mediated forms of communication do have the potential to shape the ways in which we interact both online and offline. But these shifts are not universal, some people may be more comfortable with technologically mediated communication but others still prefer real-time, face-to-face conversation. Others may reach some sort of balance between their online and offline activities. So whilst Turkle's research findings provide an important warning about technologically mediated communication, we also need to be cautious about making generalized claims about a widespread decline in embodied social interaction. Furthermore, Turkle also recognizes that we cannot simply renounce communication technologies because they permeate our lives in so many ways. Even so, I have sympathies with Turkle's conclusion since no matter how difficult, it is time to look again towards the virtues of solitude, deliberateness and living fully in the moment. I have sympathies her conclusion that 'no matter how difficult, it is time to look again towards the virtues of solitude, deliberateness, and living fully in the moment' (Turkle, 2011, 296). In other words, it is important to cultivate and reinforce our awareness of our real-time, embodied presence and ensure that this is not diminished by an over reliance on technologically mediated social relations.

In this chapter we have considered the ways in which the terms 'online' and 'offline' identities made sense when there were clear distinctions between them. Since when the World Wide Web and online environments first came into widespread usage they were accessed by desktop computers that were anchored to a specific place. This fixed placement of desktop computers seemed to reinforce the idea that accessing online worlds was something separate to everyday interaction in the world around us. From this we can begin to discern the context in which Leary, Wertheim and Franck made pronouncements in which there were clear distinctions between online and offline identities. The discussion indicated that Turkle's early work (1994) was one of the first studies to indicate that online environments offer new opportunities for developing aspects of our identities and facilitating new forms of personal expression. On the other hand, we also considered arguments that online identities are reductive in comparison to embodied interaction. Synthesizing both sides of the argument, the discussion considered the ways in which physical embodiment is crucial to online identities, such as using the sense of sight to view avatars, or the sense of touch to click a mouse and navigate through different screens.

We also considered the ways in which there are important differences between embodied markers of identity such as class, race and gender and online identities. So whilst people appear to be able to swap gender, or change race via online identities, such opportunities are not open to all members of society. For example, some people may not have the material or cultural capital to access the Internet and create online identities. Furthermore, for some members of society the embodied markers of their identity seriously impinge upon their everyday lives, such as their health and the educational and career opportunities that are available to them.

In contemporary culture, those who have access to the Internet and the necessary economic and cultural capital may mix online and offline activities on a daily basis. Through the use of smart phones and tablet computers it is now possible to simultaneously engage in online and offline activities, blurring the distinctions between them. This point was emphasized in Turkle's vignette of Pete who made references to his life mix and the mash up created by online and offline activities. Moreover, young people today have grown up alongside digital technologies and are accustomed to simultaneously engaging in online and offline activities. Particularly, for this generation it may no longer make sense to even talk about online and offline environments because they are so intertwined.

In this chapter we have also considered the ways in which impression management, especially on social networks, is becoming an important pursuit. In some ways this could be regarded as an extension of existing impression management practices whereby we stylize our physical appearance through clothing, how we speak and the objects we possess. Yet with social networks the labour involved in impression management is not immediately obvious. Choosing what photographs to upload to Facebook, composing witty tweets or editing our blog may not appear to be a form of work since it is part of our social life and interaction with friends, families and colleagues. But drawing upon Hearn's arguments, the discussion pointed out that the construction and maintenance of our profiles on social networks involves labour and in this way contributes to contemporary capitalist systems and values. For as Hearn's work indicates, digital literacy (the skills required to engage in impression management), the idea of self-presentation and the flexible personality suits the requirements of the labour market and the needs of a rapid and global networked economy.

In addition, this chapter has pointed out that class, race and gender remain important markers of our embodied social interactions and engaging in online games or social networks does not simply erase them. Furthermore, it is important that we challenge the realities of economic disparities, racial intolerance and gender inequalities through various means. In some cases, online networks and special interest groups can be a way of fuelling prejudice and intolerance amongst their members. But social networks can also be a way of promoting and drawing people together for the purposes of challenging social inequalities and actively bringing about change (Shirky, 2009). After reflecting upon the various arguments that have been made about online and offline identities, my own view is that it is vital to consider the embodied aspects of our identities. I do not consider this to be merely sentimentalism because our embodied relationships with the world around us provide the ontological grounds for our existence and a firm foundation for our ethical relationships and responsibilities to one another. Indeed, human interaction is multi-modal, involving face-to-face dialogue, touch, displaying emotion and even smell. And as yet online identities cannot replicate the full dimensions of these multi-modal forms of human interaction.

5

Virtual Reality, Transcendence and Flights from Embodiment

Visions of transcending the limitations of physical embodiment are a recurrent feature of many cinematic representations of virtual reality in films such as *Tron* (1982, dir. Steven Lisberger), *Brainstorm* (1983, dir. Douglas Trumbull), *Ghost in the Shell* (1995 dir. Mamoru Oshii) and *Source Code* (2011, dir. Duncan Jones). At first glance these films appear to present an alluring fantasy whereby technology facilitates an escape from the limitations of the physical body. Yet, as this chapter will indicate, these films raise a series of intriguing questions about the relationships between the mind, body and immersion in virtual worlds that require further scrutiny. As these films are popular forms of entertainment and have the potential to reach millions of viewers worldwide through cinematic release, DVDs and later through television broadcasts, they are important cultural resources which can have a significant impact upon our perceptions of the relationships between embodiment and technology. To limit the scope of this study of technological transcendence, I will focus attention on the film *Source Code*. One of the reasons for studying this particular film is that it presents a vision of transcendence from the limitations of embodiment through immersion in virtual reality technology and it also raises a series of questions about placing the body within a cybernetic loop with machines.

In this chapter, I will contend that the idea of technological transcendence from the physical body via immersion in virtual reality is problematic because it is based on a misleading dichotomy of mind versus body. At the same time, however, *Source Code* can offer us a useful way of working through our anxieties about embodiment and our relationship to technology.

To explore the relationships between embodiment and virtual reality in more detail, I will refer to the work of René Descartes (1596–1650). One of the reasons for doing so is that Descartes's work has been incredibly influential in Western culture for the past four-hundred years. Specifically, the chapter will engage with the distinction Descartes makes between the non-physical mind and the physical body and test out these ideas in relation to *Source Code*. On this basis, the chapter will indicate that Descartes considers the body as a mechanical, physical object whereas the mind is mysterious and beyond empirical investigation. Recent developments in the field of neuroscience, however, are starting to erode this mind and body dualism (Damasio, 2000; Rose, 2005). Neuroscience is a complex field which differs in terms of methodologies and approaches to the humanities and is not my area of expertise. Despite these shortcomings, I have drawn upon and interpreted ideas in the field of neuroscience, to the best of my ability in order to illuminate some of the issues arising from contemporary media representations of virtual reality and transcendence from embodiment. In this chapter I will suggest that despite recent advances in neuroscience and an increased understanding of cognition, there continues to be a fascination with the idea of transcending the physical mind and engaging in mental flights of fantasy. Moreover, this fascination with transcendence can be detected in contemporary media representations of virtual reality.

To further investigate the ways in which technological development intersects our understanding of the relationships between embodiment and technology, I will also refer to information theory and cybernetics from the 1940s onwards and the work of Claude Shannon (1916–2001) and Norbert Wiener (1894–1964). In particular, I will explore the ways in which Shannon's work on logical propositions and signal processing has been extended to areas such as human cognition. Therefore, it will be necessary to consider what happens when Shannon's work is removed from the context of engineering and information systems and extended to the operational features of the brain. For example, can cognitive processing be regarded as a pattern of signals that can be mapped and recreated mechanistically? So by drawing upon the work of Shannon and Wiener, I will attempt to trace the connections between the mechanistic approach to cognition and the placement of the human body within a cybernetic system and *Source Code*.

Yet, in this chapter I indicate the ways in which the interpretation of a popular film such as *Source Code* is more complex and multifaceted than the communication models that are found within information systems and engineering, for *Source Code* operates on a number of registers, including visual and acoustic special effects, mise-en-scéne, dialogue, framing and editing. Moreover, audiences do not just receive and decode signals when watching films. Instead, there are many factors that contribute to the ways in which viewers interpret films, including their prior knowledge of other films (intertextual relationships), the context in which they view the film (cinema, DVD) and their prior life experiences. Therefore, during this chapter I will highlight the ways in which *Source Code* is polysemic, offering multiple interpretations which enable viewers to interpret the film in a variety of ways. Indeed, I will contend that the film encourages viewers to imagine and reflect upon a series of intriguing questions about embodiment, subjectivity, the real and the virtual.

The narrative of *Source Code* focuses on Captain Colter Stevens (Jake Gyllenhaal) and his return from active military duty in Afghanistan. After returning to the United States, Colter is recruited to a new mission that involves the use of a simulation training program, known as Source Code. After a bomb explodes in a commuter train going into Chicago on a busy morning, Colter is placed within the Source Code simulation in order to investigate the bomber's methods and prevent any further attacks. In this way the film can be interpreted in the context of historical events such as the terrorist incidents that occurred in the United States in September 2001 and the subsequent deployment of US military and other NATO forces in Iraq and Afghanistan and heightened concerns about the possibility of further terrorist incidents on US soil.[1]

Taking a wider historical perspective, the use of simulation techniques within the US military can be traced to the development of flight simulators in the 1930s and 1940s which combined moving, mechanical mock-ups of aeroplanes with film images. By the 1980s developments in computer graphics produced virtual reality training simulations such as the work conducted by Thomas A Furness III on the Super Cockpit Program (1986–1989) at Wright Patterson Air Force Base in Ohio. Today, flight simulations and combat-based simulation programs are an important part of military training and strategy development. Taking these points into consideration, we can discern that within the context of *Source Code*, Captain Colter probably is familiar with simulation programs since it is likely that they would have formed part of his military training. Of particular relevance to this discussion is that when Colter is immersed in Source Code it appears that his mind becomes separated from his physical body. Therefore, I will now turn to the work of Descartes in order to consider the implications arising from Colter's immersive experiences in more detail.

1 Four aeroplanes were hijacked on September 11, 2001; two led to the collapse of the World Trade Centre in New York, another plane damaged the west side of the Pentagon and the fourth plane crashed en-route to Washington in Pennsylvania (source http://www.bbc.co.uk/history/events/the_september_11th_terrorist_attacks).

Cartesian dualism

To begin with, it is important to provide a brief biographical sketch of Descartes's life since this sets the context in which his work was written, particularly in relation to the tensions arising from his ideas about the mind and body. Descartes was educated at a Jesuit college in La Fléche and later studied at the University of Poitiers. After he completed his studies he travelled to Germany, Paris, and from 1629 to 1649 he lived in seclusion in Holland (Heilbron, 2003). It was during this period of seclusion, circa 1633, that he wrote *Le Monde* (also known as *Treatise on Light*), but this was not published in his lifetime. One of the reasons it was not published was that in this work Descartes defended the heliocentric view of the universe. At the time he was writing, Descartes was aware of the ways in which Galilei Galileo (1564–1642) had come under attack from the Roman Catholic Church for his views on heliocentric cosmology and stood trial for heresy (Livingstone, 2006). Galileo was found guilty, forced to recount his views and sentenced to house arrest, and it seems that Descartes wanted to avoid a similar fate. Consequently, *Le Monde* was finally published in 1664, fourteen years after Descartes's death. The tensions between Descartes's philosophical and scientific approach and the official doctrine of the Roman Catholic Church are also pertinent to the distinctions he made between the mechanical aspects of the human body and the immaterial human spirit in *Meditations* (1640). Indeed, the philosophical framework proposed by Descartes makes a clear distinction between the mind and body. This framework was constructed in the backdrop of religious beliefs about the soul and transcendence from the physical body, as in the context of the doctrines of the Roman Catholic Church the soul is thought to be eternal and released from the mortal body upon death. Yet Descartes faced a challenge to reconcile the world of matter which could be mapped and measured with the soul which is invisible and not subject to empirical testing. To overcome this challenge, Descartes contended that there were two distinct realms of existence, res-extensa (matter) and res-cogitans (mind). As we shall see, remnants of Descartes's framework can still be detected in the twenty-first century, but transcendence from embodiment is no longer regarded as a metaphysical or religious concern; instead, it is often represented as something which can be achieved through technology.

When discussing embodiment, Descartes (1985) postulates that the human body has receptors that passively respond to the environment. He illustrates this point by referring to the use of wax and a seal, an example that would be relevant to the time in which he was writing. Descartes draws attention to the ways in which wax makes an impression on a seal and claims that this is similar to the imprinting of sensory impressions on the mind. The sense impression is conveyed to a central processing area which Descartes refers to as the 'Common Sense'. The term 'Common Sense' has multiple interpretations in the context of Descartes' work since it can refer to the brain, the mind or the soul. Within Descartes's philosophical framework, the central processing area of the common sense can also retain and recall the impressions, which have been imprinted upon it, and this is one way of explaining memory.

According to Descartes, sensory perception does not involve an actual object; rather, it involves the transmission of information. So how do we make sense of the complexity of the sensory information that is available to us? Descartes states that we concentrate on specific aspects of our experiences. Specifically, he claims that we do not see things in themselves, but only an 'abbreviated' representation of them (1985, 43). If we see an abbreviated re-presentation of reality, this suggests that there is a complex mediating or translation process between the world and our sensory experiences. So our understanding of reality and the virtual cannot be based solely on the notion of unmediated or direct sensory experiences. Following Descartes's line of thought, if we see an abbreviated or mediated version of the world around us, then to what extent is it possible to unequivocally know that what we are seeing is reality? As discussed, Descartes' work provides a backdrop of uncertainty about the relationships between the mind and body, the virtual and the real, which are relevant to contemporary representations of virtual reality. And the issue of uncertainty is especially relevant to the ways in which perception and making judgements about reality are represented in *Source Code*.

In *Le Monde*, Descartes examines the difference between our sensations and the objects that produce them. The experience of seeing light involves both the eye and cognitive processes, and these are quite different from the actual physical qualities of the objects that produce the sensory experience. Expanding this line of thought, Descartes (1985) adds that words bear no resemblance to the physical world; yet we gain experience of our world through language:

> Now if words, which signify nothing except by human convention, suffice to make us think of things to which they bear no resemblance, then why could nature not also have established some sign which would make us have the sensation of light, even if the sign contained nothing in itself which is similar to the sensation. (81)

The suggestion here seems to be that nature produces signs which give the impression of light even though the object which caused it contained nothing like that sensation. Here Descartes appears to pre-empt, to a certain degree, more recent developments in neuroscience. On a similar note, Antonio Damasio states that 'the images you and I see in our minds are not facsimilies ... The object is real, the interactions are real, and the images as real as anything can be. And yet, the structure and properties in the image we end up seeing are brain constructions prompted by an object' (2000, 321). Descartes was formulating his ideas in a very different context to contemporary neuroscientists who undertake clinical and other empirical investigations to test and validate their hypotheses. Instead, he was drawing on reason and logic to construct his philosophical framework. Taking this approach Descartes aimed to increase the understanding of ontological and epistemological concerns about human existence. Although Descartes considers vision to be the noblest of the senses, he distrusts sensory perception as a means of acquiring knowledge. Overall, what we find in Descartes's framework is that sensory perception is dubious and cannot be trusted, a point that seems to undermine Colter's attempt to understand the Chicago train bomber through immersion in *Source Code*.

In *Treatise on Man* (*De Homine*) (1662) Descartes contends that external objects strike the senses and activate them and that the senses correspond to certain areas in the brain. When the senses are stimulated the nerves pull on the brain, opening up the corresponding region. According to Descartes, this process enables animal spirits to enter the brain. Furthermore, depending on how hard they are pulled different levels of animal spirit enter the brain. If the nerves are pulled too much, they can break and this is what causes pain. On the other hand, if they are pulled vigorously but do not break, they produce a sensation of pleasure. Although the idea of animal spirits or the nerves as lever-type mechanisms seems strange, from a contemporary perspective, the basic principle of sensations activating certain areas in the brain is not. For there are now numerous ways to map and record brain activity, including electroencephalography (EEG), computerized axial tomography (CAT), positron emission tomography (PET) and magnetic resonance imaging (MRI).

In a contemporary setting, one of the problems arising from the use of MRI imaging is that it can be used to bolster claims that particular regions of the brain have a causal relationship to behaviour. However, David Eagleman points out that 'imaging methods make use of highly processed blood-flow signals, which cover tens of cubic millimetres of brain tissue. In a single cubic millimetre of brain tissue, there are some one hundred million synaptic connections between neurons' (2012, 173). In other words, imaging and mapping the brain is a very detailed and complex process. To explicate this further, Eagleman writes that 'modern neuroimaging is like asking an astronaut in the space shuttle to look out the window and judge how America is doing' (2012, 173). Therefore, MRI can be used to detect some macro-level changes in neural activity, but smaller, subtler changes remain hard to discern. So, despite advances within neuroscience we are still a long way from fully understanding cognition, consciousness and our embodied interaction in the world around us. Furthermore, these limitations also have implications in terms of the idea of transcending the limitations of the physical body through sensory immersion in virtual environments.

Criticism of sensory experience as a means of acquiring knowledge also has implications for the understanding of immersion in the Source Code program and Colter's mission to investigate and locate the bomber of the Chicago-bound train. According to Descartes's framework, there are four faculties that help us investigate the world of objects; these are the intellect, imagination, sense perception and memory. But for Descartes imagination and sense perception are considered lower in rank than the intellect because they provide unverifiable knowledge. In particular, experiential knowledge is dubious because it is prone to error. On the other hand, Descartes values deduction because this involves reason and exercises the intellect, thereby providing a more precise form of knowledge. It is in this context that Descartes values mathematics as a means of acquiring knowledge, because it is primarily based upon logic and universal principles. Whilst the virtual world of Source Code is generated by algorithmic, logical principles, Colter's immersive experience and his quest to locate the bomb on the train does involve sensory perception and emotionally driven interactions in the world around him. But the knowledge acquired by Colter's sensory immersion in Source Code and events in the real world remain ambiguous. In this way, the film raises doubts about sensory experience, knowledge and making distinctions about the virtual and the real.

Descartes's philosophical investigations aimed to provide a greater understanding of human existence and our place in the world. To achieve this aim, Descartes utilized the principles of reason, logical argument and personal reflection. Ultimately, Descartes came to the conclusion that consciousness defines human existence. As science writer Margaret Wertheim (1999) puts it, Descartes's famous maxim 'I think therefore I am' meant that reality was grounded in the immaterial realm of thought rather than the physical world.

Therefore, a twofold distinction is made between an objective world which is meaningless and the mind which makes the world meaningful. The problem with this viewpoint is the ways in which it severs the connections between material, biological processes and cognitive functioning (Damasio, 2001). In addition, Laurie Spurling remarks that Cartesian dualism 'resulted in the doctrine of the "ghost in the machine", where the body was considered to be the physical container of an ethereal mind' (1977, 21). By drawing upon contemporary scientific and clinical research, Damasio takes a rather different approach, stating that 'it is possible to conceive of mind states as biological phenomena, which means that they are material, physical, and have a definable spatial and temporal extension'(2001,194). However, Damasio concedes that such a task is not easy:

> The challenge is enormous, the gap between what we know about neural patterns and what we experience as mental patterns has not been filled in, but we do not need to believe, as Descartes did, that the gap is an intransposable abyss between the material and the immaterial. (2001, 194)

Taking these points into consideration, it is necessary to be cautious when considering Descartes's philosophical postulations, since he was writing within a different context to our contemporary age. Furthermore, there continues to be critical debates and controversies relating to the implications arising from Cartesian dualism (Damasio, 2001; Kirkeboen, 2001).

Despite the problems arising from Descartes's work, the idea of mind and body separation persists within contemporary culture and can be detected in contemporary representations of virtual reality. For instance, one of the first cinematic representations of immersion in computer-generated virtual worlds was *Tron*, in which the protagonist Kevin Flynn (Jeff Bridges) transcends his physical body and is reanimated in a computer-generated game. A decade after the release of *Tron*, the relationships between sensory experience, knowledge and virtual reality were featured in the film *The Lawnmower Man* (dir. Brett Leonard, 1992). In this film, the protagonist Dr Lawrence Angelo (Peirce Brosnan) seeks to run trials to augment intelligence and recruits Jobe (Jeff Fahey), a humble gardener, to be a research participant. During the course of the project, Jobe is transformed into a post-embodied entity through immersion in virtual reality. Once immersed in virtual reality Jobe views a series of images that increases his intelligence; for example, he learns Latin in less than two hours. What is problematic is that the knowledge Jobe acquires is not the result of continual practice, reason or critical analysis. Instead, his intellectual prowess is the result of spectatorship and appears instantaneous and effortless. However, the impression of instant knowledge and intelligence is subsequently undercut when Jobe has a psychotic break. After this psychotic break, Jobe casts off his physical body by downloading his consciousness into a computer-generated world, discarding the real world in favour of the virtual.

Neuroscience

So what can contemporary developments in the field of neuroscience tell us about the relationships between the mind and body? More importantly, how can they help us elucidate the contemporary media representations of embodiment and immersion in virtual environments? Damasio contends that there are two aspects to studying consciousness. One aspect involves the study of the sensory perception of objects: how we see, hear, smell or touch something or someone. According to Damasio, this aspect of consciousness involves the creation of what he calls a mental movie in the mind, which is a series of images that are generated from neural patterns. The second aspect of consciousness concerns our sense of self because there is an observer or a witness who processes sensory perception. Although this may initially appear to be a restatement of the Cartesian dualism of mind and body, Damasio regards both aspects of consciousness as physical rather than a metaphysical phenomenon. Indeed, Damasio argues that the brain generates the mental movie in the mind and the observer. But as we have seen, even within the contemporary field of neuroscience there are still unexplained aspects of neural activity. Of note is that Damasio pays particular attention to the terms he uses to refer to the causality of neural states. Taking a cautionary approach, Damasio states that mental images *arise from* neural patterns; he does not state that *they are* neural patterns. Indeed, Damasio acknowledges that there are still gaps relating to the cause and effect of neural activity and our experience of the world around us.

Damasio divides his study of cognition into proto, core and extended consciousness. The proto self refers to basic evolutionary processes in which distinctions are made between an organism and external objects. The proto self relates to those neural patterns which regulate and maintain life and operate beneath the level of self-conscious awareness. Core consciousness relates to a basic sense of self and immediate lived experience and interaction with objects.[2] However, extended consciousness refers to a more complex sense of self-identity and extended temporality since it involves memory and personal history. Extended consciousness draws on our memories and the learning we have acquired through previous experience. In this way, extended consciousness enables us to draw upon and manipulate what we know. The ability to manipulate knowledge requires a sense of self that has a longer duration than the fleeting-like qualities of the proto and core aspects of consciousness. Cultural factors can also influence extended consciousness through the construction of values, beliefs and customs. Damasio also discusses the ways in which extended consciousness allows us to rise above our immediate concerns of survival and to establish ethical relationships with others. Applying Damasio's framework to the interpretation of *Source Code*, it is possible to argue that Colter draws upon his personal history and prior experiences in the military when he searches for the bomb on the simulated train journey into Chicago. Furthermore, as he becomes more familiar with the Source Code program through repeated immersions in this environment, he starts to remember the features within it and learns how to manipulate the knowledge he has gained from prior experience.

Cybernetics

Colter's repeated excursions within the simulated environment of Source Code can also be thought of as placing him in a cybernetic feedback loop with computers. To gain further insight into cybernetics, I will now turn to the ways in which *Source Code* relates to the work of Claude Shannon and Norbert Wiener in the field of information systems and computer science. The work of Claude Shannon in the post–Second World War era has been incredibly influential in the field of information theory, computer science and engineering. Shannon studied Electrical Engineering at the MIT and obtained a doctorate in Mathematics in 1940 (A Dictionary of Scientists, 1999). From 1941 to 1972, Shannon was employed by AT&T Bell Telephone Company and also worked as a professor of Communication Sciences and Maths at MIT. Shannon drew upon the work of George Boole, the founder of Boolean logic (Hillis, 1998). Boole had devised a system for processing statements based on logical operations (such as and, or, not and copy), and Shannon applied this system to a series of electrical switching

[2] Damasio uses the term 'object' to refer to material things or people.

circuits. Specifically, his work indicated that circuits could be open or closed, a process that corresponded to true/false in Boolean algebra. If a statement were true, then a connection would be established between switches; if false the connection would be broken. The basic principle of logical operations can also be applied to an electrical current which is divided into two states, flowing (1) or not flowing (0), and associated with something that has a true or false value. Shannon's model of communication was based on engineering principles in which there is a source (transmission) and a receiver (destination). The sender produces a signal which is transmitted to a receiver, but interference (or noise) can also occur during this process.

Shannon's model of communication and interaction is a technical one and it has useful practical application in engineering and telecommunication systems. But this model is problematic when it is applied to other areas such as human communication and the generation of meaning. On this point, Katherine N. Hayles observes that 'Shannon himself frequently cautioned that the theory was meant to apply only to certain technical situations, not to communication in general' (1999, 19). Yet, as we shall see some of Shannon's ideas about communication, logical propositions and information processing can be detected within contemporary representations of human and computer interaction in popular film and literature.

In the work of Norbert Wiener, attention is given to the ways in which we perceive the world around us via the sense organs and how this is processed by the brain and nervous system. According to Wiener, sensory perception is a process that involves information exchange, calibration and re-adjustment. Wiener creates analogies between humans and complex machines, stating that they both operate via a process in which an input of data triggers an output/action in the world. The initial output then becomes data input, providing feedback on the action that was performed. To illustrate this process, Wiener refers to the ways in which automatic doors have two different operational mechanisms to open or to close. He goes on to say that organisms are capable of overcoming entropic decay through homeostasis and complex feedback mechanisms. For instance, a healthy human body maintains a temperature of 98.6 Fahrenheit even in very different climatic conditions. So, for Wiener, the body is not a fixed object, rather there is a dynamic interplay within the body and the wider external environment. Upon close examination it appears that Wiener is making analogies at a general level and does not claim that humans are simply complex machines. Underscoring this point, Weiner states:

> When I compare the living organism with… a machine, I do not for a moment mean that the specific physical, chemical and spiritual processes of life as we ordinarily know it are the same as those life-imitating machines. I mean simply that they both can exemplify locally anti-entropic processes. (1968, 31)

Wiener also makes analogies between mechanistic operations and neural firing (or not firing) across a synapse in the human brain. But he warns that the idea that the brain is a digital machine is something that cannot be taken too literally because synaptic firing involves many different interconnections and chemical changes and the build-up of activity to a certain threshold.

Technological transcendence

Yet despite the cautions issued by Weiner and later by Hayles regarding the application of mechanistic principles to human beings and cognitive processes, these ideas have found a new mode of expression in the work of computer scientists such as Ray Kurzweil and Hans Moravec. Kurzweil is a major figure in the debates around computer technology and the implications this may have for our understanding of human life. What we find in the work of Moravec (1988) and Kurzweil (1999, 2003, 2009) is a strong sense that technology can be used to enhance, upgrade or even transcend the body. Indeed, Kurzweil claims that human beings have limited capabilities which can be overcome using technology. For instance, he compares and contrasts human neural activity with computer processing to illustrate the limitations of human cognition. Kurzweil contends that 'our thinking is extremely slow: the basic neural transactions are several million times slower than contemporary electronic circuits' (2009, 8). Kurzweil also applies terms that are usually associated with computers to human life, for instance, 'our version 1.0 biological bodies are … frail and subject to a myriad of failure modes' (2009, 9). Kurzweil goes on to argue that we need to discard human life version 1.0 and upgrade to a more sophisticated program.

Kurzweil even outlines a technological solution to human mortality, claiming that technological advances will mean that human life does not need to be valued or defined in relation to death. 'As we move towards a non-biological existence, we will gain the means of "backing ourselves up" (storing the key patterns underlying our knowledge, skills and personality), thereby eliminating most causes of death as we know it' (2009, 33). In actual fact, Kurzweil regards the mind as software which is trapped in the hardware of the body.

Consequently, when the hardware malfunctions, the software crashes and we die. Kurzweil recognizes that his scenario of non-biological life would involve updating software systems since we would not want our brains to be using out-of-date software that was incompatible with others. Talking about such issues as archival, Kurzweil claims that 'information lasts only as long as someone cares about it' (2009, 329). But how do we decide what is relevant to us at any one time? What if we discard something as irrelevant but later it becomes extremely valuable to us? Furthermore, will the individual have the freedom to decide what information they care about and want to preserve? What if they want to hold onto information which is valuable to them but not conducive to the well-being of society as a whole (such as information about how to conduct terrorist acts or build bombs)? How will we reconcile what we care about personally with societal values?

Surprisingly, media scholar David Gauntlett dismisses Kurweil's argument, stating that this could be 'just about some Californian millionaires wanting to live forever' and that 'we don't need to worry about that…' (2011, 202). Kurzweil also recognizes that there will be some who disagree with his vision of transcendence, but he caricatures this into fundamentalist and Luddite views rather than recognizing other more measured forms of opposition. For instance, he claims that Luddite views are an absolute negation of technological development in which 'humanity will be better off without change, without progress' (2009, 415). Of note is that Kurzweil's statement appears to conflate change with progress. Those who challenge technological development may not necessarily be against social, economic or scientific progress. Instead, those who challenge technological development could open up critical debates with which to consider what changes we want to make and how this can be achieved without negatively impacting upon the things we value at present and want to sustain. Such critical debates may also form the bedrock for establishing public policy, ethical or legal standards towards technological change.

Kurzweil is quick to dismiss the opposition to his vision of transcendence by stating that the advantages brought about by technology will be so good that they will appear irresistible. He then lists numerous benefits to the vision he proposes, including improved health, wealth and intelligence. He recognizes the slippery slope argument whereby small technological changes will gather momentum resulting in large-scale changes. Yet, for Kurzweil this slippery slope does not have a downward trajectory; instead, it is moving upwards towards more desirable modes of existence. Furthermore, he argues that there will be a growing acceptance of non-biological intelligence because it will become increasingly difficult to distinguish between humans and machines. Kurzweil also tackles the criticism that enhanced forms of technological intelligence will only be made available to rich and powerful members of society. According to Kurzweil, technology goes through a process whereby it is initially expensive, but then the price and availability come down and it becomes more widely accessible.

The technological vision set out by Kurzweil also needs to be contextualized in relation to contemporary concerns and challenges. For as Hayles (1999) remarks, 'in a world despoiled by overdevelopment, overpopulation, and time-release environmental poisons, it is comforting to think that physical forms can recover their pristine purity by being reconstituted as informational patterns in a multidimensional computer space' (36). Moreover, for some people human mortality and the fear of death may make the idea of transcending the human body and becoming immersed in virtual environments an attractive proposition. As discussed, Kurzweil claims that people will come to accept the copying or downloading of the brain into computers because it will vastly improve their mental capabilities, including memory, pattern recognition and reasoning. But what we find in Kurzweil's work is an overly optimistic account that denies some of the shortcomings of technology. Mapping the neural connections of the brain would provide a snapshot of someone's cognitive processes but to what extent could it really capture the totality of their existence? How would the downloaded version of the neural connections of the brain continue to learn and respond to novel situations; for instance, would this be something that was built into the computer program supporting this post-embodied life? And if such post-embodied existence becomes possible, then what happens when there is a requirement for upgrades or even threats such as system crashes or computer viruses?

It seems that remnants of Shannon's communication model and Kurzweil's vision of a technological solution to human mortality can be detected in *Source Code*. For instance, each time Colter becomes immersed in the simulation he discovers new clues resulting in an output (new knowledge/insight). This output then becomes the starting point for his next immersion and so on. The idea that communication involves the transmission and reception of signals (as per Shannon's model) can be also detected in a dialogue between Colter and his commanding officer, Sergeant Goodwin (Vera Farmiga). After Colter's first immersion in Source Code, Goodwin communicates with Colter via an intercom. Goodwin says, 'Captain Colter Stevens this is Beleaguered Castle acknowledge transmission', thereby implying that a signal is sent from a sender to a receiver. However, Colter has difficulty responding to the transmission due to the disorientation that has occurred between the simulated world of Source Code and Colter's location within the cockpit of what appears to be a Stealth bomber. Goodwin then asks him, 'do you have a visual signal'. When Colter sees Goodwin on a visual display monitor he has a sense that she is familiar, but he cannot recall her name. She replies that his memory can be enhanced by a 'thread one alpha memory pattern'. In this way, the film suggests that Colter's memory is a pattern of information that can be recorded, stored and retrieved through technological systems in the way that Kurzweil envisages.

When Colter is first immersed within Source Code he has trouble adjusting to this multi-sensory simulation. Colter's disorientation is heightened when he goes to the train's rest room and looks in the mirror, because instead of seeing his own reflection he sees the reflection of another young man. Afterwards, Colter learns that within the simulation he takes on the appearance of a teacher called Sean Fentriss. After numerous immersions in the simulation, Colter starts to find new clues about the bombing and he also learns more about his own predicament. For instance, in one scene, Colter asks his fellow passenger Christina (Michelle Monaghan), who is Sean Fentriss's girlfriend, to find out information about him. After using a mobile phone to access the World Wide Web, Christina tells him that Captain Colter died whilst on active duty. This incident raises a number of questions relating to agency, the acquisition of knowledge and the relationships between the simulation and the world beyond. For example, to what extent can the simulated Michelle decide to access the World Wide Web? Does Colter have the ability to influence or even control the behaviour of other simulated people on the train? Or, are Colter's agency and Michelle's actions simply the result of the parameters of the program itself?

After realizing that he takes on the appearance of someone else in Source Code, Colter confronts Sergeant Goodwin and demands to know more about his situation. Goodwin and her commanding officer tell him that due to technological developments it is now possible to capture and recreate the final minutes of someone's life. So the Source Code simulation is based on a recording of the last 8 minutes of Sean Fentriss's life. The idea of recording someone's life experiences and simulating them is reminiscent of some of the proclamations issued by Kurzweil and Moravec regarding the reverse engineering and downloading of consciousness. For, whilst Colter is immersed in Source Code, he can relive Sean's last 8 minutes over and over again. In this way there are similarities between *Source Code* and the film *Groundhog Day* (dir. Harold Ramis, 1993), in which Phil Connors (Bill Murray) relives the same 24 hours but acquires the ability to change what happens. However, in Colter's case his ability to take on the appearance of someone else and change the simulation also starts to unsettle our pre-existing ideas about mind and body, the virtual and the real.

Through repeated immersion in the Source Code simulation, Colter appears to go beyond the parameters of the program and learns more about the bombing and his own predicament. Using the knowledge he has gained through experiential learning in the simulation, he seems to be able to alter what has happened in the past and also what happens in the future. There are some similarities between Colter's repeated immersion and experiential learning in Source Code and playing computer games. To begin with, Colter has to explore different realms and acquire information. Through practice Colter's skill level increases, he begins to pre-empt what is going to happen in Source Code and in the end he starts to move beyond the parameters of the program. This process also appears to be incremental; for instance, in one of his earliest forays in Source Code, Colter is situated within a single railway carriage. However, in a similar manner in which game players reach different levels within a game, Colter soon begins to make forays into different parts of the train, the vestibules, other carriages and the restroom. Each time he moves through space, he also alters time by changing what has happened. After repeated immersive experiences within the train, Colter starts to extend his capabilities and exits the train at a station on the way to Chicago. Eventually, Colter tracks down the bomber to the station car park, thereby altering his spatial location and making a critical intervention in what happens to the train.

Yet the impact of Colter's interventions in Source Code on the real world is ambiguous. We see a television news report that shows the bomber's capture, but it is unclear if this footage is a simulation or what has happened in reality. Our understanding of what is simulated and what is real is destabilized further as Colter's predicament is slowly revealed. At the start of the film it appeared that when Colter was not immersed in Source Code he was located inside a Stealth bomber, but it later transpires that this is another simulation. Instead, Colter's body is dismembered and enclosed in a high-tech tomb. So in retrospect, viewers are able to piece together a backstory in which Colter was seriously injured during active combat duty in Afghanistan. Nonetheless, Captain Colter's usefulness to the military has not ended because his skills, knowledge and experience can be used to help prevent another terrorist incident.

Brain–Machine Interface

It is useful to place the representation of Colter's dismembered body, which can regain functioning through becoming part of a cybernetic feedback loop with a computer simulation, within the context of research into Brain–Machine Interfaces (BMIs) since this helps elucidate some important issues regarding the relationships between embodiment and technology.

In the late 1990s, the neurosurgeon Roy E. Bakay (M.D.) at Emory University conducted procedures on two people. One person had Amyotrophic Lateral Sclerosis (Lou Gherig's Disease) and the other had a stroke. The procedure Bakay performed involved implanting electrodes in the brains of these patients and linking them to a computer. After that, their neural signals could be transmitted to a receiver which was then used to control the cursor on a visual display. In his analysis of BMI research, Antony Bryant recounts (2010) that in 2001 researchers at the North Western University, Chicago, produced an interface that links the brain stem of a Lamprey fish with a robot. The rationale underpinning the project is that in the near future brain-controlled prosthesis could be developed for people with conditions such as Parkinson's disease. More recently, in 2012, researchers at Brown University have conducted research involving a 58-year-old woman who was paralysed by a brain stem stroke. This research involves using electrodes so that this woman's neural activities could be tracked and used to control a robotic arm. Bryant (2010) skilfully argues that research in the field of BMI is justified on the basis that it will ultimately benefit disabled people or those with serious injuries. However, this sort of research also impacts upon our understanding of human life and our relationship with machines. As mentioned earlier in relation to the work of Shannon and Wiener, whilst there may be some similarities between models of neural functioning and information, this does not necessarily mean that cognition is simply a matter of digital, binary processing. Indeed, Hayles warns that when ideas from cybernetics, information theory and engineering are abstracted and applied to human life, they can be used to 'the unwarranted conclusion that there is no essential difference between thought and code' (1999, 61). The neuroscientist Steven Rose (2005) also makes an important distinction between the digital processing of computers and cognition. Rose points out that 'information is empty without a system to interpret it, to give it meaning' (2005, 54). Whilst computers process signals according to a series of precise logical principles, human communication can involve complex modes of symbolic signification, interpretation and the generation of ambiguity. Outlining the difference between symbols and signals, Rose states:

> Symbols are conventional, resting on a shared understanding that any particular symbol is a token representing a particular feature of a particular referential class. Unlike a signal, which can be regarded as an instruction to behave in a particular way, a symbol guides not behaviour but understanding. (2005, 106)

Therefore, human communication and the generation of meaning can be regarded as a symbolic process which differs from the signals or precise instructions that underpin the operational mechanisms of computers. Taking these differences into account, it seems that neural processes can be connected to computer processing, for instance to amplify their operational mechanisms, but this does not mean that cognition as a whole is amenable or comparable to computer programming.

The complexity of human cognition in relation to computer programming is highlighted by Colter's immersion in Source Code. For it is possible that in Damasio's terms Colter's proto and core consciousness are maintained through the interventions of machines. In other words, Colter's basic life functions would not function without machines. Whilst Colter is coupled to these machines, he appears to be unconscious and in a lucid dream-like state. If Colter is unconscious, then we could interpret the Source Code program as something that is connected to his neural processes and provides him with a sense of autobiographical self, or what appears to be extended consciousness. Even so, the position is complicated further when Colter asks for the equipment that is keeping him alive to be switched off; in other words, he becomes conscious and aware of the fateful wounding of his physical body and wants to die. This raises questions about whether Colter somehow rejects the realism of the Source Code program. But what enables Colter to make the transition from believing he is either in the cockpit of a Stealth bomber, or on a train bound for Chicago, to the understanding that he is seriously injured and kept alive by machines? The precise ways in which Colter moves from an unconscious state to self-awareness is not fully explained in the film. Yet, I would contend that the ambiguity that arises from these changes to Colter's awareness is what intrigues the audience. By leaving some things unexplained, the film offers audiences an opportunity to generate their own interpretation of Colter's predicament. In this way the film seems to illustrate the differences between the precision and logic of computer programming and the rich ambiguity of the human interpretation of symbolic modes of signification.

Before the equipment that maintains Colter's life functions is switched off, he asks to be immersed in Source Code one more time. During this final immersion, the train reaches its destination and Colter and Christina walk through downtown Chicago. Colter appears to have transcended the limitations of his physical embodiment and escaped from the confined space of his high-tech tomb. In addition, he also starts to increase his agency within the Source Code program, shaping the environment to meet his desires. As they walk through the city, Colter and Christina approach *Cloud Gate*, a sculpture by Anish Kapoor (2004–2006) in Chicago's Millennium Park. Made from polished steel, the sculpture both reflects and distorts the world around it, which neatly fits with the film's premise of unsettling the relationships between the virtual and the real. In the closing moments of the film, Goodwin receives a text message from Colter, who is still immersed in Source Code. The message reveals that he is now living a post-embodied existence in the simulation. In the closing scene, we see Colter and Christina looking at *Cloud Gate*, but it is still Sean's face that looks back. In the end, *Source Code* presents us with the idea that Colter can transcend his embodied condition and take on the simulated appearance of someone else. Cognition has become a pattern or code that can be transmitted from one vessel (Colter's body) to another (to a simulated body) within a computer program.

However, by taking an intertextual approach to *Source Code*, it is also possible to see that the film does not represent radically new or subversive ideas about the relationships between embodiment and technology. For the idea of transcending the limitations of embodiment for a simulated life is a frequent trope within science-fiction film and popular literature. For example, in Philip K. Dick's novel *UBIK* (1969), there are 'half lifers' – people who have some residual brain function and are kept alive through cyrogenics. Similarly, in William Gibson's novel *Count Zero* (1986), the character Josef Virek is a billionaire who is kept alive through technological means and who makes rare personal appearances via simulated environments and tele-conferencing. The idea of recording the last moments of someone's life is also central to the narrative of *Brainstorm*. In this film, scientist Lillian Reynolds (Louise Fletcher) and her colleague Michael Brace (Christopher Walken) are working on technology that can be used to provide multi-sensory experiences. However, the US military wants to take over their project and use this technology to induce traumatic experiences as a means of torture. Before she can expose the military's intervention in this research, Reynolds suffers a heart attack and records her final moments. Subsequently, Brace is able to use virtual reality technology to play back Reynolds's recording. In *Brainstorm*, Brace appears to be driven by a spiritual quest, to learn about life and death, whereas the driving force behind Colter's immersion in Source Code is a military one, to prevent a terrorist attack on US soil.

Greg Egan's novel *Permutation City* (1994) also unpacks some of these issues surrounding a virtual afterlife. The novel's protagonist, Paul Durham, has a complete brain scan and uploads a copy of his consciousness into a virtual reality environment. The copy then develops its own experiences from the point it was uploaded, but it still has memories it shares with Paul Durham. Within this virtual world the markers of human life span, such as childhood, adolescence, adulthood and old age, no longer have any meaning. Faced with the prospect of an unlimited time span copies have to find ways of filling their time, to keep life interesting. Yet many copies find that having an unlimited time span becomes oppressive and want to terminate their existence. In this way the novel challenges us to think further about the implications arising from the idea of a technological afterlife.

As mentioned earlier, research in the field of BMIs has already resulted in coupling the brain stem of a Lamprey fish with a robot. However, the Japanese anime film *Ghost in the Shell* takes this idea further, since the brain stem of the protagonist, Major Motoko Kusanagi, is encased within a technologically advanced cyborg body. At the start of the film the Major is employed by the Japanese government in a special task force called Section 9. Of note is that her cyborg body belongs to her employer and this places limits on her individual agency. As the narrative unfolds the Major seeks to understand the nature of her existence and expand her agency. By the end of the film, the Major transcends the limitation of her government-shackled cyborg body by coupling her mind with an artificial intelligence program called the Puppet Master.[3] But as Brian Bolton observes, although Major Kusanagi can utilize the features of her cybernetic implants to extend her consciousness into computer networks, this means that her mind is vulnerable because it can be accessed by other cybernetically enhanced humans or computer programs such as viruses. We could also extend

[3] In *Ghost in the Shell* the Major's brain is represented as a hybrid consisting of a human brain stem and cybernetic implants.

these ideas to the analysis of *Source Code*; for instance, if Colter's mind can be completely immersed into a computer-generated environment, does this mean that others can access his mind too? Could the military implant false memories in Colter, and if so, how can he make distinctions between what is real and what is not?

Both *Ghost in the Shell* and *Source Code* raise a number of questions about the idea of a stable identity which is anchored to a physical body. In this way they present us with some interesting ideas about how we construct our sense of identity. According to Damasio, we generate a sort of movie in our minds which is comprised of a series of images. These images make sense because they are related to a singular self that is anchored in the body. There is stability to our inner bodily states, for 'the body must operate within a limited range of parameters if it is to survive' (Damasio, 2000, 22). Our bodies change, grow and some parts start to decay, yet the internal multi-cellular environment needs to be relatively stable. Our bodies are continually re-arranged as cells die and are replaced, but the structural design of our bodies is maintained. So for Damasio, it is our body that gives rise to a singular self. Even in the case of multiple personality disorders (also known as dissociative identity disorder) only one identity is prominent at a time. So what has happened to Colter's sense of identity as he takes on the appearance of Sean Fentriss and becomes immersed in the virtual world of the Source Code?

The singularity of our embodiment also relates to our sense of agency. Indeed, Damasio contends that agency 'requires a body acting in time and space and is meaningless without it' (145). But what happens to Colter when he lives a virtual life in the Source Code program? Will his physical body be kept alive by the military for future missions of a similar nature? If the recording of Fentriss's consciousness is changed by Colter then to what extent does Sean Fentriss still exist? Similarly, what happens to Major Kusanagi's identity after her merger with the Puppet Master? *Source Code* and *Ghost in the Shell* present us with a series of baffling questions about identity, memory and embodiment. Specifically, these films seem to be aligned with Kurzweil's claims that cognition is an informational pattern that can be transferred to different platforms, cyborg bodies (*Ghost in the Shell*) or virtual worlds (*Source Code*). However, the veracity of memories is also undercut by these films in various ways. In one scene in *Ghost in the Shell* a refuse worker talks to his colleague about his family, but later these memories are shown to be fake. Major Kusanagi witnesses the moment when the refuse worker is told by her colleagues at Section 9 that his family memories are fake and this precipitates doubts about her own memories and sense of identity. At the start of *Source Code*, Colter believes that he is in the cockpit of a Stealth bomber, then on a train going into Chicago, but later he learns that these are both simulations. Yet Colter later realizes that his identity is coupled to his sensory experiences in the Source Code simulation and that this offers a number of opportunities to him. In particular, Colter appears to be able to exercise agency, through decision making and changing the simulated environment.

Although *Source Code* is set in the present day and resonates with contemporary events such as US involvement in Afghanistan, Colter's military mission is also set in the context of Hollywood cinematic conventions of personal fulfilment through romantic involvement. Indeed, the film ends with Colter's final immersion in Source Code and moments of romantic fulfilment with Christina. At the same time, this conventional cinematic ending appears to be undercut because Christina is a simulated figure within the Source Code program. Furthermore, if Source Code is a recording of Sean Fentriss's life, then Christina's character is linked to their prior experiences together. Yet as Colter appears to exercise his agency in Source Code by making decisions, there is the possibility that he can change Christina's character through his interactions with her. In this regard *Source Code* leaves viewers with an uncertain ending. Colter's mind has become fully immersed in Source Code, but it remains unclear to what extent he is an active agent within the program.

In conclusion, this discussion of *Source Code* has drawn upon Cartesian dualism, information theory and contemporary neuroscience to explore the representation of transcendence from embodiment through immersion in computer-generated environments. As this discussion has indicated, Descartes's studies aimed to construct a solid and reliable framework for the production of knowledge. We have seen that for Descartes sensory perception was dubious and could not provide a solid foundation for knowledge. But after much deliberation Descartes arrived at his conclusion that what he could be sure about was that he was a thinking being. In other words, he could not deny or be confused about his conscious awareness of himself, which is summed up in his famous dictum: 'I think therefore I am'. We have also considered the ways in which Descartes's distinction between mind and matter relates to religious concerns about the existence of the soul. As discussed, Descartes's work has been incredibly influential in Western culture and his ideas continue to permeate contemporary representations of mind–body separation and immersion in virtual reality in popular Hollywood film. Yet what we have found through this discussion is that the idea of mind–body separation within these films appears to be connected to wish fulfilment and the desire to transcend the limitations of our embodiment through technological rather than religious means.

The discussion has also pointed out that analogies have been made between the processing of signals in the field of electronics and engineering and how neural processing operates. In this way cognition can be regarded as a physical rather than metaphysical phenomenon. But as this discussion has pointed out, for writers such as Wiener and Hayles these analogies operate at a generalized level and it is unwise to extend them too far because cognition appears to be more complex than the informational models that are used within the context of engineering. Nonetheless, it is also conceded that in some cases research in BMIs suggests that neural signals can be placed in a cybernetic loop with machines as a medical intervention to help people with serious injuries or impairments.

Whilst Colter appears to have transcended the limitations of his dismembered body, it is possible that his consciousness will still be used by the military for further missions, so his situation is not as unbounded as it initially appears. In this regard the military contexts surrounding the development of virtual reality technology are an important consideration. As this discussion has pointed out, simulation techniques are utilized by the military to help train personnel for military operations and as a means of developing scenarios and strategies. This military use of virtual reality technologies is also tied to concerns about maximizing the success of military operations and optimizing resources. The crossovers between military simulation and gaming techniques and commercially available computer games are a further point to consider in relation to *Source Code*. For as this discussion indicated, Colter's mission to find the bomb on the Chicago-bound train is reminiscent of a computer game. Indeed, it was pointed out that Colter has to move through different levels of the train and engage in experiential learning to fulfil his quest. Through repeated immersion in the simulation Colter becomes more adept at navigating through the train, responding to changes in the environment and detecting clues that will lead him to the bomb and ultimately to the capture of the bomber.

It is also possible to interpret *Source Code* as offering viewers a form of wish-fulfilment, since Colter is able to overcome his serious injuries and find personal happiness in a simulated world. In this way the film could be regarded as allaying our fears by presenting a technological solution to mortality. Furthermore, when wish fulfilment is combined with a lack of in-depth knowledge of information theory and cybernetics, it may result in the scenario presented by *Source Code* appearing credible to some viewers. From this perspective, *Source Code* can be interpreted as offering a dramatic encounter with technology which could prepare the way for the growing acceptance of human–machine interfaces and immersion in simulated environments.

Unlike the purported logical purity of information theory, *Source Code* is a polysemic text. Therefore, some viewers may reject the wish-fulfilment scenario it seems to offer. Indeed, another interpretation is that *Source Code* provides a contrast to the celebratory tones of Kurzweil and Moravec because it reveals some of the problematic aspects of post-embodied existence in virtual reality since the parameters of Colter's agency remain unclear. A further possibility is that after the closing scenes of the film when Colter and Christina walk through Chicago, his life-saving equipment is finally switched off and he is allowed to die.

6

Technology, Embodiment and Being-in-the-World

This chapter explores the relationships between technology and the phenomenological concept of being-in-the-world through a discussion of what initially appears to be two very different examples: *Osmose*, an installation art piece that can be described as a virtual work of art that was produced in the 1990s, and the film *Avatar* (2009, dir. James Cameron). *Osmose* was exhibited internationally and gained significant attention within the field of art history, thereby contributing to theoretical debates about digital art (Rutledge, 1996; Davis, 1999; McRobert, 2007). In contrast, *Avatar* is a blockbuster Hollywood film that has reached a much wider, global audience and has garnered both critical and popular attention.

This chapter indicates that there are some interesting and significant connections between *Osmose* and *Avatar*. Firstly, both *Osmose* and *Avatar* draw upon the experience and imagery of immersion in oceanic space when representing technological forms of immersion. Secondly, the archetypal image of the tree is given particular prominence in both *Osmose* and *Avatar* as an ecological and existential motif. Thirdly, this chapter will indicate that *Osmose* and *Avatar* utilize technology to highlight important questions about the impact technology has upon our lives and the world around us. Therefore, in this chapter I will explore the argument that whilst *Osmose* and *Avatar* may aim to evoke nature, these works are produced using an array of technological devices, which are part of the global nexus of production and consumption which impacts the environment through the usage and transformation of natural resources. At the same time, however, I will indicate that *Osmose* and *Avatar* could be catalysts for different ways of thinking about technology, embodiment and wider environmental concerns.

In this chapter, I will use the term 'technology' to refer to the ways in which human beings fashion and shape the environment through material and symbolic practices. In this sense, rather than referring specifically to computers or the World Wide Web, this chapter acknowledges that technology is embedded within different social structures and cultural practices and has a long and complex history. In the second decade of the twenty-first century technology permeates our lives through the shaping and organization of agricultural practices (genetically modified crops), mediating communication (satellite television, digital audio broadcast, social networking) and transportation (air-traffic control), to give just a few examples. Taking such factors into account, I will contend that *Osmose* and *Avatar* take us beyond a simplistic polemic of nature versus culture because our social, economic and cultural lives have become thoroughly intertwined with technology. Therefore, it is no longer possible to turn away from technological development and replace it with an idyllic, bucolic vision of nature.

In this chapter I will also investigate how *Osmose* and *Avatar* invoke some of the ideas and concepts surrounding technology that are found in the work of Martin Heidegger (1889–1976). During Heidegger's lifetime there was an intense period of extensive technological change such as the development of photography, film, radio, television and telecommunications. Yet he also lived through a period in which there were tremendous atrocities such as the horrors that took place in Auschwitz and the atomic bombs that destroyed Hiroshima and Nagasaki. In 1933, Heidegger joined the Nazi party, one explanation given for this is that he saw this as a way to overcome the nihilism that he thought threatened Western civilization (Calhoun, 2002). Another view is that Heidegger was critical of the technological mind-set which he associated with the United States and USSR (Inwood, 2005). After the end of the Second World War Heidegger was called to account for his association with the Nazis and prevented from teaching until 1951, and arguably this aspect of his life and work is still problematic today (Inwood, 2005; Buchanan, 2010). Even though there are problematic aspects to Heidegger's work, he does offer an insightful critique of a voracious and controlling technological mind-set and the impact this has upon our relationships with one another and the world around us.

Of particular relevance to this chapter is Heidegger's essay *The Question Concerning Technology* (1954), in which he explores the question of being (human) and our relationship to Being (the ground of existence).[1] In this essay,

[1] Heidegger's essay stems from a lecture he gave to the Bremen Club in December 1949 called 'The Enframing'.

Heidegger asserts that technology has a long historical trajectory and links this line of argument to epistemological and ontological concerns, particularly in relation to manipulating and mastering the world around us. So, for Heidegger 'technology is neither neutral nor instrumental' (Botha, 2003, 159). Therefore, Heidegger is not simply dismissive of technology, rather he is concerned about the ways in which technology is 'a particular mode of disclosure' (Botha, 2003, 159). In this regard, one of the main themes in Heidegger's essay is that technology becomes problematic when it is driven by attempts to dominate and master Being. Catherine Botha explains that 'Heidegger's fear is that someday, calculative thinking would be accepted and practised as the only way of thinking. Calculative thinking (*Rechnendes Denken*) is the type of thought that deals only with the quantifiable and the measurable' (2003, 163). There is nothing wrong with calculation per se, since it can be an extremely useful. However, problems occur when calculation becomes a dominant mind-set that occludes other ways of thinking. For as Heidegger cautions, we may become so enamoured and caught up with this mind-set that we do not realize where it is taking us or what we might lose in the process.

Heidegger contends that technology can become coupled with a mind-set in which the world around us is regarded as a set of resources or a standing reserve (*Bestand*) to be exploited for our own ends. In many ways this viewpoint on technology can be set in the context of European thought during the past 400–500 years. Yet, non-Western traditions offer a different way of thinking about technology (Parkes, 2003). For instance, Graham Parkes draws scholarly attention to the ways in which Heidegger is thought to have engaged with Daoist classic literature in which the term *wuwei* refers to activities that flow with the unfolding of natural processes. In contrast, *youwei* refers to activities that disrupt natural forces. Parkes points out that technology can be considered on a spectrum between *wuwei* at one end and *youwei* at the other. For example, the term *wuwei* could be used to refer to 'Windmills, sailboats, watermills… implements that make use of the natural forces of wind, water, and gravity without abusing or using them up', whereas *youwei* would be technologies such as nuclear power or genetically modified crops (Parkes, 2003, 29).

A particularly useful way of thinking about technology can be found in Heidegger's term *aition* 'to which something else is indebted' (1993, 314). In particular, Heidegger's work emphasizes interdependence as many different elements come together and unfold in the natural world. For instance, sunlight, air and gravity play a part in plant, animal and human life. Therefore, human beings do not stand above nature; rather, human life is part of an interdependent web of existence. According to Heidegger, acknowledging this interdependence generates a sense of reverence for the forces and relations that exist in the natural world. This is in sharp contrast to the view that the natural world is a resource to be mastered, manipulated and in some cases exploited by human beings for our own ends. Arguably, echoes of Heidegger's line of argument can be found in the work of American biologist Edward O. Wilson and his conception of two major opposing forces and worldviews surrounding consumer society: exemptionalism and environmentalism. Wilson (1997) identifies exemptionalism with arguments that human beings somehow stand above nature and that we can utilize technological developments to overcome ecological constraints. In contrast, environmentalism places human beings within an interdependent relationship with the natural world. From this perspective long-term sustainable stewardship is the way to maintain the web of life in which humans are one strand, amongst many.

Taking these points into consideration, it is not that technology per se is problematic. However, as Heidegger contends, technology is problematic when it is coupled with an overarching emphasis on acquisition, domination and commercial exploitation. In this regard the relationships between Being and human being and the concepts of *wuwei* and *youwei* provide a useful way of interpreting *Osmose* and *Avatar* and to open up debates about technology, change and development.

Osmose

Osmose is an installation artwork that consists of a series of computer-generated spaces that are accessed by an individual user wearing a specialized head-mounted display and cyber-vest. By wearing this technological apparel the movement of the user's diaphragm is monitored in order to provide interactivity within and between these different computer-generated spaces. *Osmose* consists of several different levels: a central clearing, tree, leaf, life world and code world. What is significant about code world is that it highlights the computer code that generates virtual environments, whereas commercially available photo-realistic computer games tend to erase their means of production. Code world features the 20,000 lines of code, including quotations from Heidegger, Austrian poet and novelist Ranier Maria Rilke (1875–1926) and the French philosopher Gaston Bachelard (1884–1962). *Osmose* also featured sampled sounds of male and female voices, but these voices were altered so the sounds were evocative rather than literal renditions of words or songs. *Osmose* is a French word which refers to that which flows between. In this way Davies's work can be regarded as a way of encouraging a series of flows between different computer-generated spaces. Of note is that whilst *Osmose* attempts to evoke nature, this virtual world is devoid of human or animal life. In this regard it is a virtual world that invites individual self-reflection and contemplation rather than offering an opportunity to interact with other subjects (human or animal).

When Davies and her team produced *Osmose*, the dominant design paradigm for mainstream computer graphics was photorealism and linear perspective. In addition, the hand-held interfaces that were used with computer games were usually based on pointing, clicking and shooting objects. In contrast, what we find with *Osmose* is that the body becomes an interface with which to explore different virtual environments. The computer graphics used to generate *Osmose* were also utilized in such a way that they produce a sense of ambiguity rather than a photorealistic simulation of nature. The use of loose strokes and flecks of colour in *Osmose* also produces an interesting contrast to the more solid, hard-edged objects that can be found in popular computer games. On this point, Beryl Graham remarks:

> Davies has started to challenge the frenetic moral vacuum of most VR 'games' with a landscape of shimmering aesthetic and perceptual impact, which has stimulated both ecstatic reactions and hot debate concerning representations of nature, or artists' access to technology. (1996,07)

The conceptual work for the design of *Osmose* began as early as autumn 1993 and was completed six months later (Porter, 1996). During this time Char Davies was director of Visual Research at a computer graphics company called Softimage, which was based in Montreal, Canada. In the early days of the company there were just four employees, but soon it expanded to an employee base of over two hundred people. Softimage became a highly successful company producing special effects for major Hollywood films such as *Terminator 2* (dir. James Cameron, 1991), *Jurassic Park* (dir. Steven Spielberg, 1993) and *Independence Day* (dir. Roland Emmerich,1996). But in 1994 Microsoft acquired Softimage and soon after that Davies left the company.[2] Notably, *Osmose* is a collective production rather than the result of

[2] Softimage has now become a division of Avid Technology.

the work of a single artist, since it was produced in conjunction with a range of specialists including Georges Mauro (computer graphics), John Harrison (software), Dorota Blaszczak (sound and programming) and Rick Bidlack (composition and programming). Through working for Softimage, Davies and her team were able to access high-specification, costly equipment such as the Onyx engine produced by Silicon Graphics. On this basis it can be contended that Davies and her team were in a somewhat privileged position because they gained access to equipment that was associated with the most cutting-edge computer graphics of the early 1990s.

During the 1990s, *Osmose* was exhibited at the Barbican in London, Museum of Monterrey, Mexico, and the National Gallery of Canada in Ottawa. In this respect *Osmose* has a history which includes where it was produced, exhibited and catalogued. Yet the work itself is a simulation which can be shown at different locations without loss of quality because there is not an original work of art that is subsequently reproduced. Moreover, as Steven Holtzman remarks, digital works of art like *Osmose* are made from bits of information, patterns of zeros and ones and 'in the digital domain, a copy is always perfect' (1997,123). On the other hand, *Osmose* cannot be reproduced in its entirety through two-dimensional media such as print, or photography. In his essay *The Work of Art in the Age of Mechanical Reproduction* (1936), Walter Benjamin (1892–1940) discusses the aura of a work of art as a singular object in time and space and how this changed as mechanical reproduction enabled many copies of an art object to be reproduced, thereby reaching a wider audience. Although Benjamin's essay was written decades before digital technologies such as computer graphics came into widespread use, his arguments about the implications of the mechanical age remain important because they also provide a useful way of thinking about art in the digital age. *Osmose* is a digital work of art that generates a certain aura when it is located in a museum or gallery since such venues are not usually a part of our everyday lives. Therefore, a visit to a gallery or museum takes people beyond their prosaic activities to a place that is specifically devoted to showcasing works of art, or artefacts. At the same time however, the aura that is associated with the gallery or museum setting is undercut in the case of *Osmose* because this is a digital artwork that can be reproduced many times over, in different locations, thereby unsettling the concept of the unique artwork.

Offering insight into the issues surrounding art and digital modes of production, Mark Hansen (2004) suggests that as art becomes dematerialized the relationship between embodiment and aesthetic experience becomes more important: 'Simply put, as media lose their material specificity, the body takes on a more prominent function as a selective processor of information' (2004, 22). In this way, the uniqueness in time and space that Benjamin talked of in relation to the aura of the work of art becomes something that is associated with embodiment. Hansen's main point is that we do not simply perceive virtual reality environments; we experience them through the body. Crucially for Hansen, perception is not disembodied. Instead, the complexity and messiness of the body come into play when we experience the world around us. Therefore, Hansen's arguments relating to embodiment are particularly pertinent to an interactive installation piece such as *Osmose*.

Davies uses the term 'immersants' to refer to the individuals who engaged with *Osmose* (Davies, 1999). According to Davies, many of those who were immersed in *Osmose* felt as if they were dreaming. 'Others felt like angels and said they were no longer afraid of death' (1999, 161). Some of those who were immersed in *Osmose* also expressed feelings of euphoria and wonder, 'while a few wept upon emerging' (161). On a similar note, Virginia Rutledge states that many of those immersed in *Osmose* 'did report feelings of tranquillity and a deepened sense of connection to their bodies, to an inner self, nature, or the cosmos, variously' (1996, 39). Yet, it is the comments made by Erik Davis (1999) that are the most intriguing since he points to the strange combination of heightened embodiment and transcendence reported by those who were immersed in *Osmose*. Davis recounts that '*Osmose* conjures up the perceptual high of a walk in the woods; many "immersants" feel at once immaterial and embodied, like angels moving with animal grace' (1999, 56). But to what extent could immersants feel immaterial and embodied? Whilst they were intimately connected to *Osmose* through the bio-feedback mechanism produced by the cyber-vest they were wearing and the rhythm of the breath, they were navigating through a landscape with no physical solidity. In this way, immersion in *Osmose* can be interpreted as an invitation to transcend the world around us for a virtual world in which gravity and the physical constraints of our bodies are inconsequential. Indeed, *Osmose* was designed in such a way that the movement of the immersant's breath supposedly produced the impression of effortless movement through a translucent landscape. Drawing upon her experiences of scuba diving, Davies attempted to create an immersive experience that was similar to floating in oceanic space (McRobert, 2007). In this way the physical weight of the immersant's body seems to be transcended as they float through different parts of *Osmose*. In 1996, I visited the exhibition Serious Games, which was held at the Laing Gallery in Newcastle upon Tyne, and experienced being-in-the-virtual world of *Osmose*. However, I found that the head-mounted display and cyber-vest tended to disrupt my overall experience of being immersed in this virtual world. In fact, after a few minutes of becoming immersed in *Osmose*, I became increasingly aware of the weight of the technological paraphernalia that I was wearing. So, rather than feeling a sense of transcendence from physicality, or floating through an immaterial space, I felt constrained by the operational features of this installation piece.

Whilst the experience of immersion in *Osmose* attempts to evoke oceanic space, this installation piece can also be interpreted as providing commentary on contemporary ecological concerns. A key feature within *Osmose* is a tree which appears in a central clearing area. The figure of the tree could generate a host of associations, including our memories of trees in the physical world and representations of trees in other paintings and drawings. The seasons and geographic location of trees also have a bearing on the way in which they are perceived. A cherry blossom in Japan during the springtime is quite different to a pine tree seen on a cold December day in the north of England. Trees are also perceived differently in terms of functionality; put crudely, for a timber merchant they are commodities whereas an environmentalist would consider them to be vital natural resources.

But the figure of the tree in *Osmose* is not associated with a specific geographical location; instead, it could be interpreted as an aesthetic or poetic signifier. As Gaston Bachelard puts it, the forest can be a poetic symbol that evokes spatial and psychological experiences:

> We might make a detailed examination of what is meant by the immensity of the forest. For this 'immensity' originates in a body of impressions which, in reality, have little in connection with geographical information. We do not have to be long in the woods to experience the always rather anxious impression of 'going deeper and deeper' into a limitless world. (1994, 185)

In this statement Bachelard indicates that being in a forest is a way of experiencing the vastness of nature in comparison to human life. In this way the forest can remind us that human beings are one strand in a multifaceted, interdependent web of life.

The references to Heidegger's work in *Osmose* are also significant because they concern the relationships between technology and being-in-the-world. Heidegger claims that art is a way of safeguarding Being, of caring for the ontological grounds that make all life possible, because it involves a different mode of thinking that takes us beyond the rational or calculative. It is in this context that philosophy scholar David Farrell-Krell writes that for Heidegger it is 'not the political but the poetical…' that will enable us to deal with the questions concerning technology (1993, 310). Explicating his argument, Heidegger turns to the differences between art, handicraft and other forms of poiēsis and modern technology. Setting out his position in the following statement, Heidegger claims that:

> modern technology does not unfold into a bringing-forth in the sense of poiēsis. The revealing that rules in modern technology is a challenging [Herausfordern], which puts to nature the unreasonable demand that it supply energy which can be extracted and stored. (1993, 320)

Taking Heidegger's comments into consideration, *Osmose* could be regarded as bringing forth an evocative experience of nature using contemporary technology rather than traditional handicraft. At the same time, however, there are tensions between a technologically produced work of art and the quest to provide a poetic experience of the natural world. In this regard, it is also important to set *Osmose* within the context of other artworks that featured computer-based technology during the 1990s. Indeed, during this period several artists were exploring the creative potential of interactivity, computer graphics and virtual reality. For instance, other works produced in this period include *Legible City* (1991) by Jeffrey Shaw, *Passage Sets/One Pulls Pivots at the Tip of the Tongue* (1994–1995) by Bill Seaman, *Placeholder* (1994) by Brenda Laurel and Rachel Strickland and *Machines for Evolving* (1996) by Nell Tenhaaf. These artists are not producing works that involve the hand fashioning of physical materials such as wood, metal or clay. Even so, they are producing works of art using computer technology to fashion new modes of engagement that can provoke important questions and help us reflect upon the interconnections between embodiment and technology. Roy Ascott, professor of Technoetic Art at the University of Plymouth, contends that a technoetic aesthetic combines technology and mind and that this 'not only enables us to explore consciousness in new ways but may lead to…new forms of art, new qualities of mind, and new constructions of reality' (1999, 66). In addition, Ascott calls for a more interactive and intimate relationship between viewers and virtual images in order to bring about a shift in consciousness. According to Ascott, 'the quality of intimacy in the relationship between artist, system and user-participant is one of the greatest importance if a technologically based practice is to engage or transform our field of consciousness' (1999, 70).[3]

[3] Professor Roy Ascott was the director of research for Char Davies's doctoral work and has written extensively on digital art.

So to what extent does a technological work of art such as *Osmose* have the potential to provoke shifts in consciousness and raise our awareness of the relationships between technology and embodiment? Ironically, one of the drawbacks of a technological artwork such as *Osmose* is technological change and obsolescence. Virtual reality and computer graphics are developing rapidly with increased processing speeds and methods of storing digital images (through the use of image compression systems), and there is often some sort of novelty value attached to artworks produced by 'cutting-edge technology'. At the same time, however, difficulties can arise as computer graphics improve to such an extent that older forms of that technology begin to look dated. Furthermore, as Alison Abbott remarks, 'digital artworks are vulnerable to extinction as the operating systems on which they are based become redundant' (2004, 17). Therefore, in the future the array of technological equipment that was used to make *Osmose* may become obsolescent. Indeed, Char Davies's website reveals that since its initial production on high-specification, expensive equipment *Osmose* has now been transposed to a personal computer-based platform.[4]

Another line of argument, however, is that technological obsolescence is an important aspect of a virtual work of art such as *Osmose*. In this sense, obsolescence is important because it highlights impermanence. Producing a work of art that is susceptible to technological obsolescence could also be a way of provoking questions about the ways in which technological development impacts what we consider to be the natural environment. In this way, the shimmering images of nature that feature in *Osmose* may help us think further about the transformation and usage of natural resources and the devastating impact this can have on the environment. Taking these points into consideration, it is possible to contend that *Osmose* highlights Heidegger's critical position regarding a technological mind-set which disrupts the relationship between Being and human being. Furthermore, *Osmose* could be considered as something which is aligned to Heidegger's concept of poiēsis by using art as a catalyst for critical reflection upon technology.

In sum, *Osmose* is an evocative virtual work of art that aims to provoke new ways of thinking about embodiment and our relationship with the world around us. Notably, the tracking of the immersant's diaphragm through the cyber-vest connected the process of inhaling and exhaling with a computer-generated

[4] http://www.immersence.com

world. By focusing on the interactive process that takes place as we inhale and exhale, *Osmose* seems to take us beyond the idea that there are solid boundaries between the subject (human body) and object (the physical world around us). Instead, *Osmose* draws attention to the ways in which breathing is a continual process of interaction between us and the environment.

In an interview with Nick Bailey (2012), Davies provides critical commentary on the ways in which our daily lives are mediated through screen-based technology in the second decade of the twenty-first century. In particular, she is concerned about the ways in which we are entranced and distracted by things such as online social networking whilst the world around us is changing as habitats become inhospitable and other life forms are in danger of becoming extinct. Furthermore, Davies asserts that environmental issues do not appear to be a pressing concern for many people (the populace, the media or mainstream politicians), and on this basis she feels ashamed to be human. As we shall see, her remark about feeling ashamed to be human is also pertinent to the interpretation of *Avatar*.

Avatar

In contrast to the small-scale artistic production of *Osmose*, *Avatar* is what David Bordwell (2006) terms 'a megapicture'. Indeed, *Avatar* is a large, multi-national cinematic production by 20th Century Fox Film. *Avatar* is set in the year 2154, at a time when the rich natural resources on Earth have been depleted. Therefore, humans have investigated opportunities to obtain resources from other sources, including the planet Pandora. The protagonist in *Avatar* is Jake Sully (Sam Worthington), a former marine who was seriously injured in action which resulted in the loss of his legs. Whilst Jake grapples with life on Earth, his twin brother Tommy is involved in a research project on Pandora, studying the indigenous Na'vi people. To do so, Tommy and the rest of the research team engage in anthropological studies using avatars, which are technological bodies that are produced from a combination of human and Na'vi genetic material. When Tommy is killed unexpectedly, Jake is offered the opportunity to take his place on the research team since he is a genetic match for Tommy's avatar. At first Jake is motivated by an opportunity to earn money through working on the research project which he can then use to have surgery to restore the use of his legs. Economic factors also play a part in the decision to employ Jake, since using a genetically identical twin saves the production cost of another avatar. But to begin with, Jake is considered by other members of the research team as a liability because of his lack of research experience and is perceived as having limited intellectual capabilities. For instance, the head of the research team, Grace Augustine (Sigourney Weaver), makes several jibes about Jake's intellectual prowess, such as referring to him as 'jarhead', a term that evokes an empty vessel.

Avatar reached large audiences through general release in cinemas, multiplexes and subsequent DVD and Blu Ray products. The Hollywood studio system tends to promote a small number of blockbusters each year, particularly around the summer holiday period and the Christmas–New Year period. Following this convention, *Avatar* was released during the Christmas and New Year period 2009. When the film was released, press reports tended to focus on the ways in which it utilized sophisticated 3D computer graphics and its enormous box office takings (Adams, 2009; Bannerman, 2009; French, 2009). In many ways the sense of realism and immersion that is proffered by 3D technology can be a way for the film industry to remain economically viable in a contemporary media landscape that is beset by threats from computer games, DVDs, film streaming via the Internet and practices such as illegal downloading of films. Taken in this context, in this chapter *Avatar* will be regarded as a major Hollywood studio production which is enmeshed within the profit-driven operational imperatives of global capitalism. Indeed, contemporary computer graphics and special effects can function as a bedazzling spectacle to maximize audiences and revenue for mainstream Hollywood films. Furthermore, these spectacular techniques can be a way of foregrounding visceral responses rather than intellectual engagement with the narrative or with problematic issues such as the relationships between technology, embodiment and the world around us.

The cinematic spectacle

Film is a medium that has been associated with spectacle, offering audiences visual stimulation and pleasurable viewing experiences since its inception (Darley, 2000). But what constitutes a cinematic spectacle changes over time because what is considered eye-catching and innovative, at a certain time, can soon become dated. The spectacle as a cinematic phenomenon is also associated with immediacy of sensation, emotion and arousal and therefore is particularly prevalent in certain genres such as horror, action and sci-fi rather than drama or romantic comedy. Darley argues that there were a spate of spectacular movies from the 1980s onwards which were driven by special effects and the aesthetics of verisimilitude. Furthermore, Darley states that in these films 'the technology itself is the message' (2000, 53). Arguably, the practice of promoting films with expensive and elaborate special effects has continued since the 1980s; for example, recent blockbuster movies released by major studios include *The Avengers* (2012, dir. Joss Whedon), *Prometheus* (2012, dir. Ridley Scott) and *Life of Pi* (2013, dir. Ang Lee). Yet it is important to question whether these blockbuster movies bedazzle us with sophisticated computer-generated imagery or encourage us to critically consider important issues such as the technological transformation of our world.

One way to approach blockbuster movies and their dazzling special effects is to turn to the work of Guy Debord and his polemical account of *The Society of the Spectacle* (*La Société du Spectacle*). Debord's critique of what he called the spectacular society was published in France in 1967 in the context of student unrest in Paris. Ten years later the book was translated into English and an updated edition was published in 1994. In the opening pages of the updated edition, in a statement dated 30 June 1992, Debord reiterates that the claims he made in the first edition are still relevant. Therefore, I will make use of Debord's concept of the spectacle to see whether this can provide useful insight into a contemporary film such as *Avatar*.

Debord's *Society of the Spectacle* can be placed in a Marxist theoretical tradition in which there is an emphasis on material production, social and economic inequalities. A key tenet within the Marxist critique of capitalism is that consciousness is grounded in specific material conditions and relationships. Yet the forces and relations of production in society are mystified through ideological processes that support the interests of the ruling class. It is from the perspective of Marxist theory that we can interpret Debord's argument that the complex division of labour and increasing commodification of social relations generates a spectacular society. On this point, Debord writes:

> Understood in its totality, the spectacle is both the outcome and the goal of the dominant mode of production. It is not something added to the real world – not a decorative element, so to speak. On the contrary, it is the very heart of society's real unreality. In all its specific manifestations-news or propaganda, advertising or the actual consumption of entertainment – the spectacle epitomizes the prevailing model of social life. (1994, 13)

One of the functions of the spectacle is to stimulate consumer desire and consumption and perpetuate the interests of the ruling, dominant class, who control the means of production. Yet the social relations behind the array of spectacular images we encounter in everyday life are obscured. Debord goes on to say that we are seduced by the spectacular images we encounter in consumer-driven capitalist culture and lose touch with our authentic needs. Debord declares that 'the whole life of those societies in which modern conditions of production prevail presents itself as an immense accumulation of spectacles. All that once was directly lived has become mere representation' (1994, 12). In this way, Debord draws upon the Marxist concept of ideology as something that mystifies social relations, distorts consciousness and our awareness of the real, material conditions of our existence.

> The spectator's alienation from and submission to the contemplated object (which is the outcome of his unthinking activity) works like this: the more he contemplates, the less he lives, the more readily he recognises his own needs in the images of need proposed by the dominant system, the less he understands his own existence and his own desires. (Debord, 1994, 23)

Upon reflection, one of the problematic aspects to the *Society of the Spectacle* thesis is that it presents us with an account of a passive spectator who is captured and entranced by manipulative imagery. Consequently, the spectator is denuded about the social and economic conditions that limit their lives and therefore they are unlikely to challenge the status quo. So what we are left with is an account of the disconnected, individual spectator who is cut off from the rest of society, passively soaking up fantasy images without challenging them. Writing about the consequences of Debord's argument, cultural scholar Richard Kaplan claims:

> In these media-packaged and corporate-supplied depictions of the good life, all the social attributes actually denied to the general populace – independent power, freedom, social connection and meaningful social action – are repackaged as 'consumer choice', or as features of the lives of celebrities in Hollywood and Washington DC suitable for vicarious consumption. (2012, 461)

In agreement with Kaplan, this view of the spectacular society and entranced, passive spectators is reductive. Furthermore, the work of the researchers associated with the Birmingham Centre for Contemporary Cultural Studies (BCCCS) such as David Morley and Charlotte Brunsdon (2005) indicates that spectators do not passively accept the images produced by the mass media. Instead, television programmes, film or advertising posters can be thought of as polysemic texts with multiple and even contradictory interpretations.

According to Debord, the ways in which we use the term 'culture' is now associated with the separation of artistic production from communities and their integration within the social sphere. For Debord the separation of community-based social relationships and culture results in the loss of an active, stimulating and provocative culture. During the 1950s and 1960s Debord was an independent scholar, freelance journalist and leader of the French group of social theorists known as the Situationists (Buchanan, 2010). One of the driving impulses of the Situationists was to abolish the distinction between life and art. Debord also promoted the idea of détournement which involves subverting and changing the meaning of existing forms and practices through quotation, collage and assemblage. In contrast, 'recuperation' is a term to describe the ways in which a subversive strategy becomes incorporated into mainstream cultural and social practices (Hussey, 2002). So, instead of having a situation whereby passive spectators soak up images of freedom and wealth but are powerless in terms of their social and economic position, Debord calls for a communist society which is built upon purposeful and meaningful human labour, community and collective political activism. However, Kaplan (2012) claims that Debord exaggerates the extent to which the spectator is manipulated by spectacular images and idealizes collective, revolutionary praxis. Kaplan writes that in Debord's work the 'romantic idea of the collective revolutionary subject' is juxtaposed with 'an image of the populace of contemporary mass society as completely dependent and manipulated' (2012, 458).

Although Debord's arguments in *Society of the Spectacle* have been influential within cultural and media studies, Erlend Lavik (2008) contends that other writers have extracted elements from Debord's work and produced generalized, abstract discussions. Lavik explains that spectacle comes from *specere* (Latin for 'to see') and refers to something on display or eye-catching. Whilst the spectacle may be visually impressive and generate a sense of wonderment from a critical and intellectual perspective, this is problematic. But Lavik contends that other writers interpret Debord's concept of the spectacle for their own purposes and apply it to very different situations. For instance, Lavik states that Debord's account of the spectacle has been used as a metaphor for commodity capitalism, social and economic relations since the Second World War. Lavik also argues that the generalized conception of the spectacle is not much use when analysing a specific film. He also claims that other writers invoke Debord as a form of name-dropping and as a form of intellectual posturing. According to Lavik, this trivializes Debord's work and tears it from its original context, which was

Marxist revolutionary politics. Consequently, Lavik states that he does not think that 'a theory as monolithic and all-embracing as Debord's lends itself to the study of film' (2008, 171). In short, Lavik argues that as a metaphor of society the spectacle is too fuzzy and imprecise. Taking these criticisms into account, we need to be wary of how Debord's conception of the spectacular society impacts agency and spectatorship. We may also need to think carefully about how applicable this theoretical model is to a film such as *Avatar* and begin to explore other ways of interpreting this film.

Whilst *Avatar* garnered interest on the basis of the use of spectacular special effects and computer graphics, in some cases the narrative of the film was considered to be conventional fare. For instance, film critic Mark Kermode, in his BBC Radio 5 live review of *Avatar*, described the narrative as hackneyed.[5] Thomas Elsaesser (2011) presents another viewpoint towards the spectacular aspects of *Avatar* in which imprecision takes on a different function. Elsaesser argues that *Avatar* is artfully ambiguous so that it appeals to a wide audience and that this is a strategy to maximize box office returns. In this regard, Elsaesser's argument resonates with Bordwell's study of network narratives in Hollywood film in which the use of multiple strands of meaning can be seen as a way of attracting a wide audience. Indeed, through the use of carefully crafted network narratives different viewers can find something of interest and engagement in popular film. From this perspective, *Avatar* gains attention from different viewers because it features provocative and controversial themes, without being too disturbing. Picking up on Elsaesser's point, the narrative of *Avatar* raises a series of questions about control, power, exploitation, colonization and invasion. For instance, the ecological overtones and anti-military stance can be seen as a way of generating interest in the film. Yet if *Avatar* was shot in documentary style, or was overtly political, this would not necessarily make the film more popular with mainstream audiences. Elsaesser suggests that the ambiguity in *Avatar* is intentional and that it is a cunning strategy employed by James Cameron and powerful members within the Hollywood film industry to maximize interest and revenue for the film. For example, Elsaesser claims that the 'planned degree of pluralism of signs' is a carefully honed strategy that aims to maximize revenues (2011, 252). *Avatar* is not alone in pursuing this strategy, says Elsaesser, since other Hollywood films draw upon myths, legends, signs and symbols from different cultures to broaden

5 Mark Kermode's review of *Avatar* was aired on BBC Radio 5 on 18th December 2009 (www.bbc.co.uk/5live/film).

audience appeal, a point that brings us back to Debord's concept of recuperation. Elsaesser also argues that the anti-military aspects within *Avatar* could be regarded in the context of the United States' involvement in Afghanistan and Iraq. Explicating this position further, Elsaesser states:

> *Avatar's* anti-Americanism is thus just explicit enough in order to flatter Hollywood's vast international market, while not too offensive for Americans of the relevant demographic to feel repelled or insulted by it. (2011, 253)

In this way, anti-Americanism becomes something that actually supports the economic interests and cultural dominance of the Hollywood film industry. After examining critical reviews of the film, Elsaesser says that criticisms relating to colonialism and double standards are correct at one level, but miss the point, that this is an intentional strategy. In particular, these criticisms

> overlook the fact that Cameron systematically planned and provoked this false consciousness at all levels; making it the very principle of the film's construction because at the allegorical level – these are the objective conditions under which the USA maintains both its military supremacy and its cultural hegemony, the two locking together not (only) by reinforcing each other, but also by openly contradicting each other. (2011, 254)

What I find troubling with this line of argument is the idea that James Cameron alone could exercise so much control regarding the production, reception and interpretation of *Avatar*. Although Cameron may have had an overall vision for *Avatar* and was actively engaged in the actual filming and direction of the film, the end result depended upon a vast array of other social relationships within the Hollywood studio system. Indeed, different companies and specialists were employed to work on different aspects of *Avatar*, including the US-based Stan Winston Studio, Industrial Light and Magic and Weta (a New Zealand-based visual effects studio). Therefore, *Avatar* goes beyond the parameters of individual authorship or production. On the other hand, artfully constructed network narratives do play an important part in the interpretative process since they invite different audiences to interpret the film in different ways according to their own beliefs, aesthetic tastes and the contingencies of their own life histories.

To explore the narrative of *Avatar* in more detail, I will now refer to the work of Joseph Campbell (1904–1987) and his discussion of myth. One of the reasons for my interest in Campbell is that his work concerning mythological tales has been used throughout history and in different cultures as a means of psychological growth and greater awakening through 'difficult thresholds of transformation' (Campbell, 1993, 10). Therefore, when examined in the context of Campbell's mythological framework, *Avatar* cannot just be dismissed as a clichéd tale; instead it can be interpreted as a contemporary rendition of the archetypal hero's journey. Indeed, this discussion indicates that there are parallels between *Avatar* and the ways in which the hero's journey is a process of transformation that involves self-annihilation and awakening to higher stages of consciousness, knowledge and wisdom.

Campbell's study of myth provides a useful analytical tool when considering *Avatar*, because he outlines the ways in which greater stages of consciousness and heightened awareness are woven into the narrative of the hero's journey. As such, we can interpret *Avatar* through Jake's awakening to heightened consciousness, knowledge and wisdom. In Campbell's terms, the hero's journey begins with a call to adventure which can take many forms such as an accident, dream or premonition. In the following statement, Campbell outlines the ways in which the call to adventure takes the hero from familiar surroundings to another realm which can be represented:

> As a distant land, a forest, a kingdom underground, beneath the waves, or above the sky, a secret island, lofty mountaintop, or profound dream state; but it is always a place of strangely fluid and polymorphous beings, unimaginable torments, superhuman deeds, and impossible delights. (1993, 58)

In *Avatar* the planet Pandora is represented as a fabulous terrain of visual delights including magnificent mountains, lush landscapes, super-luminous plants, exotic animals and lithesome indigenous people known as the Na'vi. In Ancient Greek mythology, Pandora was created by Zeus to punish mankind for stealing fire from Prometheus. A central feature of the story of Pandora is that she removed a stopper from a jar, which unleashed evil into the world (Roberts, 2007). From a mythological perspective, Pandora is also associated with the ways in which humans have shaped the world around them (through the use of fire), and this line of thinking is also relevant to the interpretation of *Avatar*. For in *Avatar* the planet Pandora provides a tableau of visual delights which is threatened by the economic driving force of corporate capitalism and the controlling and covetous technological mind-set of those who support this agenda.

As Campbell recounts, one of the first parts of the hero's journey is the crossing of a threshold, a process that involves danger and enticement. When Jake first uses his avatar to explore the Pandoran landscape, he becomes separated from the other members of the research team. Alone, he encounters a dazzling array of different flora and fauna, including superluminal plants. In this regard *Avatar* seems to recreate some of the natural wonders that can be found in the ocean and Cameron's experience of deep-sea diving.[6] So when viewers see the amazing superluminal plants on Pandora, this can be regarded as an invitation to reflect upon the wonders of the natural world on our own planet. On this basis, parallels can be drawn between *Osmose* and *Avatar* since they both seem to represent aspects of the natural world, especially oceanic space, through technological means.

After exploring the landscape of Pandora Jake meets Neytiri (Zoe Saldana), who is the daughter of a Na'vi Chief, and she begins to instruct him in the different dimensions of Na'vi culture. Jake learns that the Na'vi have an empathic connection with the environment and with their animal companions. For example, the Na'vi have a bonding ritual in which their braids are used to connect to companion animals. Lisa Sideris observes that 'with this bonding, a kind of neural interface occurs between the Na'vi rider and his or her steed; the rider now feels what the animal feels and synchronizes perfectly with its movements and thoughts' (2010, 470). Central to the Pandoran landscape is the Tree of Souls, which holds the collective memories of the Na'vi and their ancestors. Indeed, the Tree of Souls marks the place where the Na'vi make important collective decisions through contacting their ancestors to draw upon their wisdom. Therefore, what we find in both *Osmose* and *Avatar* is that the figure of the tree is connected to the idea of being rooted to the environment and to a complex set of interdependent relationships.

In Campbell's terms, Jake's separation from the research team and meeting of Neytiri marks the crossing of a threshold and the unravelling of his former identity. After crossing the threshold Jake undertakes a series of trials and ordeals which help him reach a greater understanding of the Na'vi and their predicament. But in the end Jake must go through what Campbell refers to as 'the belly of the whale', a process which involves self-annihilation. By passing through the belly of the whale Jake dies to his old self, his human body, and can be re-born.

As discussed, Na'vi culture is based on the principle that they are one form of life within a greater ontological field of what Heidegger terms 'Being'. But the central premise of the film narrative is that Being is threatened by human activity. Heidegger's view of technology and the concept of the standing reserve are

[6] James Cameron's experience of deep-sea diving is outlined in the website http://www.deepseachallenge.com

particularly pertinent to *Avatar* since the film concerns the corporate extraction of a substance known as unobtanium from Pandora. In this way, unobtanium is a resource to be extracted and utilized for economic gain, regardless of the environmental destruction this will cause. However, the very name of the substance suggests that it is unobtainable and that the quest to exploit natural resources for commercial gain will be thwarted.

Heidegger illustrates the difference between respecting and exploiting nature by referring to modern forms of mining and traditional forms of agriculture, and these issues are also pertinent to *Avatar*. According to Heidegger, the rationale of modern forms of mining is to extract resources and stockpile them, whereas in traditional forms of agriculture the soil is cultivated, tended and valued for the bounty it produces (under ideal conditions). But as Heidegger remarks, 'agriculture is now the mechanised food industry' (1993, 320), so it is important not to get side-tracked into a nostalgic view of agriculture versus mining or to simply polarize the relationships between nature and technology. However, the representation of unobtanium in *Avatar* can also be interpreted according to contemporary concerns about the relationships between the mining of minerals in places such as the Congo and the production of technological devices such as computers and mobile telephones. For instance, the connections between the supply and demand for minerals (such as tungsten, colten and cassiterite), the mobile phone industry, child labour and the funding of militia in the Congo are highlighted in a documentary film *Blood in the Mobile* by Frank Piaseki Poulsen (2010).

In contrast to modern technology Heidegger contends that authentic craft (technē) is a process in which human beings and their tools have equal involvement in the production of the final product. In this way production involves poiēsis because it is a process of bringing something forth which is based on gratitude and indebtedness. Yet, according to Heidegger this practice of creativity and bringing forth has been replaced by a process of enframing in which tasks are objectified. Arising out of these conditions is a form of production which is driven by stockpiling or in Heidegger's terms producing standing reserves for human use.

Heidegger contends that enframing not only changes production processes; it also changes the status of human beings, for humans also become a standing reserve, a resource that is underpinned by the principle of maximized efficiency at minimum cost. This places humanity on a particular trajectory which Heidegger terms 'destining' (*geschick*). As we become enframed by technology we lose our connection to Being, but we do not even realize this due to the pervasiveness of technology in everyday life. Ultimately, Heidegger is of the view that humans have dethroned Being and that the world has become a standing reserve because it is regarded as something instrumental that serves our needs and purposes. Yet we may be so embroiled within technological enframent that we do not regard this as a loss; instead we consider the ability to increasingly control Being as something which is advantageous to us. At present our environment is shaped and intertwined with various forms of technology from agricultural processes to micro-electronics. Consequently, it is becoming increasingly difficult to untangle the relationships between technology and the natural environment, as plants and animals are now monitored through technological tagging, data collection and information processing. This process is exemplified in a report by Dan Simmons for the BBC programme *Click* (2012) about the Extreme Project in Brazil, which highlights the ways in which technology intersects formerly remote regions of the world. Specifically, Simmons's report indicates that the project involves the establishment of mobile communication technology (the use of 3G signals) in the Brazilian Amazon. This enables local schools to use laptop computers to access the World Wide Web. So it is very difficult to uphold distinctive categories such as technology and the environment. Instead, in contemporary life there is a complex set of interactions between these categories.

On the other hand, in many ways the representation of the plight of the Na'vi and the threat of environmental destruction on Pandora in *Avatar* can be interpreted as a rallying cry to protect indigenous ways of life in different parts of the world. For instance, Alexi Barrionuevo (2011) has investigated the ways in which indigenous communities have protested about the construction of the Belo Monte Hydroelectric Dam in the Amazon. In particular, the organization Amazon Watch raised concerns about the environmental impact of the dam, because this construction project would mean that some land areas would be flooded whilst other parts of the Xingu river would dry up, which would impact upon fishing and the indigenous people's way of life. Of note here is that Barrionuevo reports that James Cameron attended a meeting with indigenous group leaders in a village alongside the Xingu River in 2010. Therefore, whilst Cameron is the director of a major Hollywood studio film production, he is also actively engaged in environmental concerns. But some reviews of *Avatar* have criticized the ecological strand to the film narrative; for instance, Philip French (2009) describes it as a 'smug sermon' whilst Guy Adams (2009) claims it is a 3D spectacle representing 'tree hugging aliens'.

At first glance it may appear that *Avatar* represents a dichotomous view of the Na'vi as an indigenous culture that reveres and is connected to nature and humans who aggressively control and exploit natural resources. But upon closer inspect this dichotomous view is riddled with contradictions. For, when Jake experiences a greater closeness to Being by venturing into the landscape of Pandora and empathizes with the plight of the Na'vi, this is achieved through the technological body of his avatar. As Stephen Baxter observes, 'To some extent Jake is like a player of a virtual reality (VR) system with the "game" being Pandora…' (2012, 244). Without his technological avatar Jake seems to be limited by his human body. This is emphasized in the scenes where he is represented as cocooned within a technological pod whilst connected to his avatar. As the film progresses Jake begins to spend more time with the Na'vi than with his human counterparts. Consequently, the Na'vi world becomes more real to him and his human life becomes more dreamlike. Jake begins to forgo the requirements of his human body, such as washing, grooming and eating in favour for spending more time using his avatar. Through spending more time utilizing his avatar and being-in-the-Na'vi-world Jake begins to alter his sense of embodiment and way of thinking.

Avatar also indicates that there are dissenting voices in terms of the relationships between different human beings and different members of the Na'vi. For instance, Grace Augustine and her research team are placed within a fraught set of relationships with representatives from the mining corporation who are responsible for extracting unobtanium on Pandora. Augustine and her team seem to represent values and beliefs about understanding other cultures by seeking diplomatic solutions to difficulties, whereas members of the corporation are portrayed as representatives of corporate greed. Furthermore, members of the military unit on Pandora are represented as brutal, lacking sensitivity or empathy with the Na'vi. In fact, from the perspective of the military, the Na'vi way of life is something to be eradicated if necessary, in order to achieve the bounty of unobtanium. In turn, there is also dissent amongst the Na'vi about how to relate to human beings and protect the rich natural resources of Pandora.

Jake can be regarded as our on-screen avatar and we are invited to identify with his actions, rejecting the controlling and voracious impulses of corporate capitalism and increasing our knowledge and appreciation for indigenous ways of life. Through a process of self-annihilation in which he transcends the limitations of his human body and egocentric self, Jake is not only able to empathize with the Other – he becomes the Other. In this way, Jake becomes part of the collective Na'vi attempt to overcome the aggressive might of the military. But to what extent can *Avatar* influence our understanding of thorny issues such as the relationships between technology and embodiment, or environmental concerns?

Saving Pandora – saving the world

In August 2010, *Entertainment Weekly* published an interview with James Cameron. The interviewer Benjamin Svetkey asked Cameron about reports concerning some audience members experiencing depression after watching *Avatar* because the world around them did not appear to be as vibrant and alluring as Pandora. Cameron appeared to brush off the remarks, saying that they were sensationalized by the media. Cameron is reported as saying that those audience members who were depressed after watching *Avatar* could go out walking in the woods, or snorkelling, in other words to get in touch with nature. From this standpoint, like *Osmose*, *Avatar* can be regarded as a film that represents the wondrous flora and fauna of Pandora in order to remind us of the magnificence of nature on Earth. However, our experience of nature is increasingly mediated in various ways. For instance, our experience of nature may be mediated through popular television series, such as those presented by Sir David Attenborough from *Life on Earth* (1979) to the more recent series *Africa* (2013) for the BBC. What constitutes the natural world is also increasingly enmeshed with culture and technology, through the digital mapping of space (Google maps), or by recording devices that are used to capture images of wildlife in their natural habitat by organizations such as Conservation International.

However, Matthew Holtmeier (2010) contends that some audience members preferred going back to the movie theatre again to immerse themselves in the fictional world of Pandora rather than engaging with the complexities of the social and political activism surrounding environmental concerns. Holtmeier examines some of the accounts of depression which were associated with *Avatar* through online discussion forums (such as www.Avatar-Forums.com). Some of these discussion threads were subsequently picked up by CNN online in the article 'Audiences Experience "Avatar" Blues'.[7] In addition, Holtmeier analysed media coverage of responses to *Avatar*, finding that a major theme amongst them was that the real world appeared lacklustre compared to Pandora.

[7] http://www.cnn.co.uk/2010/SHOWBIZ/Movies/01/11/avatar.movie.blues

A common solution to post-Avatar depression in the online forums was to go see the film again in order to retreat from the real world and become immersed in the fictional one. On the other hand, Holtmeier also found that the film did raise the profile of the plight of indigenous people and could be used as a platform for taking action in the world around us. In doing so, Holtmeier claims that *Avatar* has raised ecological concerns even amongst those with no prior commitment, or involvement in such issues.

Yet multi-national media corporations, such as Rupert Murdoch's News Corporation, of which Twentieth-Century Fox is a subsidiary, do wield incredible power in terms of providing information and shaping public opinion. Indeed, our knowledge and understanding of important issues such as climate change and environmental concerns are mediated through the discourses we find in various aspects of the media, in newspapers, television and radio. Therefore, some spectators may be bedazzled by 3D imagery in *Avatar* and give scant attention to the other aspects of the film such as environmental themes or concerns about the rights of indigenous people. Furthermore, as Mark Bartlett points out, *Avatar*

> [r]equires that we face the horribly discomforting fact of our complicity in the continuing global genocide of indigenous peoples (in China, the Americas, India, Brazil, Palestine, and other places) and in the ecocide of nonhuman biological life. (2012, 289)

On this basis, *Avatar* cannot be dismissed as just a populist fantasy with a hackneyed plot. Instead, *Avatar* can be regarded as a film that provokes some uncomfortable questions about the ways in which the impulses of corporate capitalism contribute to the depletion of natural resources and degeneration of indigenous ways of life on Earth.

To summarize, this chapter has indicated that *Osmose* and *Avatar* take us beyond reductive dualistic arguments of nature versus technology; instead they utilize technology to raise questions about technology itself. As we have seen, in the case of *Osmose*, immersants are invited into direction participation with a virtual world through the use of a cyber-vest. In this way the living, breathing body is intimately connected to technology. However, since the 1990s, when *Osmose* was produced, further developments have taken place in terms of human–computer interfaces. For instance, there are now devices which connect embodiment more intimately with games and other media such as Microsoft Sidewinder or the Nintendo Wii.

In this chapter, we surveyed *Osmose* from a historical perspective and found that it employed diaphanous imagery rather than the solid objects more commonly found in computer games during the 1990s. From this perspective *Osmose* was considered to be ground-breaking and exemplary. Indeed, from the perspective of the mid-1990s the computer graphics which generated the virtual world of *Osmose* were possibly as cutting-edge as *Avatar* seems to contemporary audiences.

Extending the discussion of cutting-edge technology and computer graphics, this chapter went on to consider arguments surrounding Debord's concept of the Society of the Spectacle. As discussed, there are some elements of what Debord terms 'recuperation' in *Avatar*, since the film draws upon myth and legends. For instance, we saw that Campbell's study of the hero's journey resonates with some aspects of *Avatar*. However, some of the initiation rites that Campbell studies in his work were extremely brutal and involved a great deal of suffering. Whereas in *Avatar*, the protagonist Jake does go through some life-threatening situations, but he seems to quickly overcome them and achieve personal happiness and fulfilment. In this way it seems that the audiences are presented with some mild jeopardy to keep them interested in the film narrative, but not too much brutality because this would be distressing.

With *Avatar* we find that mythological themes and imagery are combined with contemporary imaging techniques which generate a series of visual delights, including lush landscapes, luminous plant life and the richly coloured adornment of the Na'vi people through the use of pigment and jewellery. By doing so, *Avatar* seems to offer audiences breathtaking, luscious images of Pandora to encourage us to care about this world and to feel a sense of anger at the thought of its destruction by human activity. Therefore, in this chapter, I have illustrated how the film resonates with Heidegger's arguments about enframing and the concept of the world as a standing reserve. Specifically, *Avatar* can be interpreted as a warning about the dangers of rapacious impulses and the rampant consumption of precious natural resources without giving due care to long-term sustainability. In this regard the point made by Char Davies that sometimes she feels ashamed to be human has similarities to the ways in which audiences are encouraged to identify with the Na'vi and their plight in *Avatar*. One troubling aspect to *Avatar* is that it suggests that diplomatic solutions to ecological concerns are ineffective. Grace Augustine and her research team go to some lengths to conduct ethnographic research, studying different aspects of Na'vi culture with a view to reaching a diplomatic agreement in relation to the human interventions which are taking place on Pandora. However, this diplomatic approach is represented as too slow and ineffective because the corporation responsible for mining on Pandora is working to commercial goals with a strict timetable. When the diplomatic approach fails, the military is brought in to enact power over the Na'vi and their world through technological means, including the use of fighter aircraft, computer navigation systems and machines. Yet, *Avatar* suggests that the courage and ingenuity of the Na'vi enables them to utilize bows, arrows and their companion animals to defeat their opponents. On the other hand, this representation of the Na'vi victory may leave us with a sense that this is a form of wish fulfilment.

In this chapter we have also considered the ways in which the use of 3D imagery in *Avatar* also helps to stimulate interest in cinema at a time when the film industry is threatened by other media forms such as computer games and practices such as the illegal downloading of films. Although Debord's arguments about spectacular bedazzling images resonate with some aspects of *Avatar*, we have seen that his position on spectatorship is problematic. Spectators do not just passively and uncritically soak up spectacular images that mask the actual conditions of their existence. At the same time this discussion of *Avatar* recognizes that there are tensions between watching a highly technological film and the idea of reconnecting to Being.

It is also important to reflect upon how *Osmose* or *Avatar* can take us beyond individual, heightened awareness and towards practical social and political activism. As noted in this chapter, James Cameron has travelled to the Amazon and met with indigenous people and heard their concerns about the environmental impact of technological developments such as the Belo Monte Hydroelectric Dam. However, in reality, the relationships between indigenous cultures, corporate interests and technological development are complex and there are no easy solutions.

In conclusion, this chapter has explored a series of questions about the relationships between technology, global capitalism, cultural production and consumption. It may appear churlish to criticize the technological wonders that are found in contemporary society, especially through works such as *Osmose* and *Avatar*, which are incredibly visually engaging. It may even seem sentimental or anachronistic to raise questions about the relationships between nature and technology. Those young people who were born in the 1990s who are referred to as digital natives (Prensky, 2001) may not be so concerned about distinctions between technology and nature, the virtual and the real. But in this chapter, I have contended that technology is not problematic per se. It is when technology is used to support the domination and exploitation of natural resources that it becomes problematic. Contemporary technologies (from mining equipment to computerized forms of communication) can be regarded as extending, expanding and accelerating the plundering of natural resources. This state of affairs is further exacerbated when natural resources become commodities to be exchanged and consumed, often for short-term gain rather than long-term sustainability.

Yet there is no going back to a pre-technological idyll; instead, technology could be used to track human activity and our impact upon the world around us and develop innovative ways of obtaining resources that have long-term sustainability. Even so, questions remain as to the compatibility of global capitalism and long-term sustainability. The political and economic arena is not the only site where the demands of capitalism and ecological concerns are played out, for culture is also the site where we can explore and debate the meaning and value of natural resources. In the end, *Osmose* and *Avatar* can remind us of our embodied connection to the world around us and the implications arising from our interaction with nature and with one another.

7

Virtuality and Simulacra in William Gibson's *Zero History*

In this chapter, I will contend that *Zero History* (2010) by William Gibson opens up a series of concerns about the relationships between signification, virtuality and reality in contemporary culture that require further investigation. Specifically, this chapter will indicate that *Zero History* raises questions regarding the rapid circulation and re-combination of signs in contemporary culture and how this process can occlude the material conditions in which signs are produced, interpreted and understood. In recent years, Gibson has moved away from the science-fiction themes of his earlier works such as *Neuromancer* (1984), *Count Zero* (1986) and *Virtual Light* (1993) and towards contemporary fiction. For example, in his more recent novels such as *Spook Country* (2008) and *Zero History* (2010) Gibson draws upon genres such as the thriller and the detective novel, often utilizing the narrative device of the McGuffin, a puzzle that at first glance must be solved but is actually a device to bring the different characters together (Knowles, 2005). What is noteworthy is that whilst some of the major themes and ideas within *Neuromancer* have received a great deal of critical scrutiny, to date, Gibson's later novels have received less attention. Therefore, in order to distinguish this chapter from other critical commentaries on Gibson's early work and to extend the scholarship on his oeuvre, this chapter focuses on *Zero History*.

The narrative of *Zero History* centres upon the investigation of a mysterious fashion brand called Gabriel Hounds and is set mainly in London in the aftermath of the global economic crisis of 2008. To make the discussion manageable, I will focus on three main areas in relation to Gibson's novel. Firstly, the chapter will trace a historical arc that begins in the early 1980s, when Gibson's first novel *Neuromancer* was published, to the present day. It was during the 1980s and early 1990s that Gibson's work became associated with the literary and cinematic genre cyberpunk which represented futuristic, technologically saturated environments and immersive virtual realities. It was also in *Neuromancer* that Gibson first used the term 'cyberspace' to refer to virtual, computer-generated space. However, this chapter will indicate that from the perspective of the second decade of the twenty-first century, 'cyberspace' and 'cyberpunk' are no longer considered terms that refer to cutting-edge technological concerns since they have been overshadowed by more recent debates surrounding New Media or Web 2.0. Indeed, the chapter will indicate that the exotic, virtual realm of cyberspace as the site of complete sensory immersion which is found in *Neuromancer* has given way to a rather more commonplace vision whereby virtuality has seeped into the interstices of contemporary life. In part, this can be regarded as a shift from immersive virtual reality to a situation whereby digital technologies intersect and overlay our experiences in the world around us. Yet this chapter will argue that whilst computer technology may appear ubiquitous and a prosaic part of everyday life, *Zero History* alerts us to the ways in which important shifts have occurred in symbolic systems of signification. For as *Zero History* indicates, digital technologies enable signs from different time periods and cultures to be re-appropriated and circulated at great speed.

Secondly, specific attention will be given to passages in *Zero History* that refer to the concept of the simulacrum, which is a sign that gains meaning through reference to another sign rather than a tangible, material referent that exists beyond the signifying system. In this way the simulacrum instigates a situation in which meaning skims across the surface of different signs within a self-referential signifying system. Indeed, *Zero History* points to the ways in which fashion and design are signifying codes that circulate in a rapid self-referential global system of brand names and imagery. Yet the novel also shows that this rapid circulation of signs can obscure the material conditions in which signs are produced, circulated and exchanged. Expanding on the debates surrounding signification, this chapter will also highlight the intertextual dimensions to Gibson's work. For example, *Zero History* will be shown to be replete with simulacra in terms of characters and objects that refer back to other novels within Gibson's oeuvre. Within postmodern theory, there is the argument that we are living in a culture of virtuality as signs have become increasingly ludic, de-stabilized and free-floating (Baudrillard, 1983, 1988; Poster 1990). Yet as I will point out in this chapter, this idea of free-floating signs has arisen from specific social, economic conditions (Jameson, 1984; Dunst 2008). Taking these points into consideration, *Zero History* cannot be regarded as simply a reflection of contemporary life; yet, it is not completely divorced from reality either. Instead, this chapter indicates that there is a complex mediating process between contemporary life in the second decade of the twenty-first century and how this is represented in literary form in Gibson's novel.

Thirdly, the chapter will indicate that *Zero History* can be a way of drawing attention to multi-faceted levels of interpretation within fashion, design and literary expression and how this differs from the operational parameters of computer code. For computer code is a signifying system that strives to achieve a fundamental correspondence between signs and meaning. In computer code ambiguity is something to be eradicated since this can lead to system errors. As Daniel W Hillis remarks, 'unlike human language there is an exact meaning in… programming language for every word and sentence' (1998, 39). In contrast to the logical precision required for computer code, the creative realms of fashion or literature thrive on ambiguity and the twisting and turning of structures of meaning. So, although Gibson's work is often considered to be technologically prophetic, especially in relation to computers, this chapter will indicate that *Zero History* points towards the importance of human agency and creativity. In order to critically analyse the representation of signs and meanings in *Zero History* in more detail, I will draw upon the work of Katherine N. Hayles (2005), paying particular attention to the distinction she makes between the operational parameters of computer code and literary expression.

Taken overall, this chapter treats *Zero History* as a starting point for considering the extent to which there is a complex, fractured set of relationships between signs and reality in the second decade of the twenty-first century. To begin with, this chapter maps out the social, economic and cultural context in which *Neuromancer* and *Zero History* were written. This will provide the context for the subsequent analysis of the shifts that have occurred, from Gibson's early representation of virtual reality in *Neuromancer* to the destabilization of signification in *Zero History*. To do so, we shall begin by placing Gibson's earlier work within the context of science fiction, cyberpunk and cyberspace.

Cyberspace and cyberpunk

In the early 1980s, Gibson published two short stories for *Omni* Magazine, *Johnny Mnemonic* (1981) and *Burning Chrome* (1982), which were followed by his first major novel *Neuromancer*, which was published in 1984. In *Neuromancer*,

Gibson used the term 'cyberspace' to refer to a virtual space generated by computer technology. At the time, films such as *Tron* and *War Games* (1983) represented computer-generated spaces as the new frontier, where adventures and conflicts took place. Indeed, during the 1980s cyberspace was attention-grabbing because it was a space in which characters could escape the world around them, transcend the limitations of embodiment and become completely immersed within it.

During the early 1980s computers were not yet fully enmeshed within our daily lives to the extent that they are today. At that time computers were mainly expensive corporate mainframe machines that were located in entire rooms because they were so cumbersome. In addition, the Internet as a way of linking computers at different locations was not yet commercially available and was mainly used by the US military and to join researchers working at major universities in the United States. At the same time, there were some inklings of how computers could develop; for instance, computer-based games in arcades were becoming popular during this period. The first computers for personal use in the home were also becoming available, such as the ZX Spectrum (1982), produced by Sinclair Research, and the Apple Macintosh (1984). Despite the advances that were taking place in terms of the technological development and commercial availability of computers, *Neuromancer* presents us with a rather ambivalent vision of a technologically saturated world. In fact, *Neuromancer* features a complex series of power struggles involving global corporations, artificial intelligence systems and hackers.

In 1983 the term 'cyberpunk' was used by Bruce Bethke as the title for a short story which was based on a small group of teenage hackers. In an introductory section to the online version of the story, Bethke states that cyberpunk is a way of imaginatively combining punk as a sub-cultural movement with technological developments in computing. The punk element in cyberpunk signifies the edginess of street culture and sub-cultural groups such as hackers. Often within the mainstream media, hackers are represented as criminals who are a menace and threat to society. Yet in cyberpunk novels and films, hackers are often celebrated and their skills revered (Levy, 2001). The term 'cyberpunk' has also been extended to other science-fiction writers, including Bruce Sterling (1986) and Neal Stephenson (1993), who were also exploring the concept of virtual worlds before the Internet and World Wide Web became widely available. Therefore, when they were writing these novels the idea of immersion in virtual worlds was strikingly imaginative. On this point, Sadie Plant aptly remarks:

> When the first of the cyberpunk novels, William Gibson's *Neuromancer* was published in 1984, the cyberspace it described was neither an actually existing plane, nor a zone plucked out of the thin airs of myth and fantasy. It was a virtual reality which was itself increasingly real. Personal computers were becoming as ubiquitous as telephones, military simulation technologies and telecommunications networks were known to be highly sophisticated. (1997, 12)

In addition, Philip Hayward and Tana Wollen remarks that science fiction contributes to the ways in which technological developments such as virtual reality are understood. So, literary and cinematic representations of virtual reality 'are significant because they have shaped both consumer desire and the perceptions and the agenda of the medium's developers' (1993, 182). Therefore, some of the ideas that feature in cyberpunk fiction can also be regarded as a catalyst for research and development in virtual reality technology. Bill Gates, CEO of Microsoft, for instance, states that 'the best descriptions of VR actually come from cyberpunk science-fiction like William Gibson's' (1996, 151).

The value accorded to *Neuromancer* within the genre of cyberpunk is also related to Gibson's literary significance. On this point, Istvan Ciscsery-Ronay remarks:

> William Gibson's career and reputation threaten to imitate the panic narrative logic of his own fictions. Gibson was immediately cited as a form of postmodern apotheosis, on the basis of a few stories and a first novel. (1992, 221)

To a certain extent the attention given to *Neuromancer* supports the interests of the publishing industry and the authorial function. As Joseph Nazare remarks, 'cyberpunk is often synecdochized as a single, token text: William Gibson's *Neuromancer*' (2003, 383). Certainly, *Neuromancer* has been incredibly influential and the scenarios it outlines are echoed in the work of other writers such as Neal Stephenson's *Snow Crash* (1993), major Hollywood films such as *Strange Days* and *The Matrix* (1999) and the Japanese anime film *Ghost in the Shell*.

In her in-depth study of science-fiction publishing and *Neuromancer*, Sarah Brouilette carefully outlines the context in which Gibson's novel was produced, and this provides us with the starting point with which to consider how Gibson's work has changed since then. Brouilette argues that 'the general championing of Gibson's text is intimately related to its situation within a particular print environment' (2002, 188). Specifically, Brouilette points out that 'the 1970s and early 1980s prove themselves an era of special concern, as the science-fiction field in general experienced unprecedented growth…' (190). But by the 1990s the publishing industry started to change in part due to takeovers which contributed to the establishment of global media conglomerates and the growing importance of chain stores for books such as Barnes and Noble, Borders and Waterstones. So by the 1990s there was an emphasis on authors who were well established and had a solid track record of commercial success. There was also an emphasis on material that would appeal to a mass-market rather than risky or edgier material by unknown writers. The changes that took place in the publishing industry placed science-fiction writers in a somewhat awkward situation, for '[s]cience-fiction publishing has had a special relationship with such corporatization, given what its practitioners and proponents depict as their status of producers of a subcultural or "ghetto" literature that is of interest only to a particular segment of the reading public' (189). From Brouilette's account we can see some of the tensions between the sub-cultural edginess of science-fiction literature and the publishing industry at large.

Signifying systems

In his study of consumer society, Jean Baudrillard claims that even sub-cultural groups are caught up in the logic of signifying systems. So for Baudrillard, sub-cultural groups offer signs of difference without posing a threat to the system itself. Indeed, he goes as far as saying that counterculture is a crucial part of the overall system of meaning because our society is based upon consumption and denunciation (1988). Taking Baudrillard's argument into consideration, it is interesting to consider the ways in which Gibson has gone from a young writer within the sub-culture of science fiction to a more mature, well-established writer of contemporary fiction. In this context, there are tensions between the ways in which Gibson's later novels may be interpreted as a critique of commodification when they are also absorbed within the global, capitalist imperatives that underpin the publishing industry. At the same time, the critical and popular attention that is given to *Neuromancer* is starting to overshadow Gibson's later novels such as *Zero History*. But upon close reading, *Zero History* is notable and worthy of considerable critical attention because it invites us to reflect on the relationships between technological development, signification and reality. Indeed, through questioning these relationships in a subtle rather than overt way, *Zero History* seems to echo the ways in which virtuality has seeped into the cornices of our everyday lives.

Baudrillard's work chiefly interrogates capitalism as a symbolic system, though he does decry 'its instantaneous cruelty, its incomprehensible ferocity [and] its fundamental immorality' (1983, 29). As we shall see, this focus on symbolic systems of signification makes Baudrillard's work particularly relevant to the analysis of signs and codes in *Zero History*. Also of relevance to the discussion is the way in which Fredric Jameson uses schizophrenia as a cultural metaphor in order to explore the ramifications of the destabilization of signs and meaning, particularly in relation to subjectivity and temporality. Alexander Dunst remarks that Jameson's concept of the subject is pessimistic, 'posting a loss of autonomy, historical understanding and intellectual capacity' (2008, 109). Talking about Jameson's usage of schizophrenia and temporality, Dunst remarks that it results in a situation whereby the world is experienced as disconnected signifiers and that this process features in 'the paranoid texts of cyberpunk and the postmodern thriller' (110). Yet, *Zero History* also raises a series of issues about the ways in which signs are removed from one historical context to another. Specifically, Gibson's novel points to the ways in which this process of historical abstraction generates an increased emphasis on the appearance of the sign and occludes the material conditions in which they are produced, circulated and understood.

It was during an interview with Timothy Leary in the 1990s that Gibson stated that *Neuromancer* is 'about the present. It's not really about an imagined future. It's a way of trying to come to terms with the awe and terror inspired in me by the world in which we live' (Gibson in Leary, 1994, 23). Whilst Gibson's statement was made in the mid-1990s, his conception of a new kind of virtual space remains influential in the second decade of the twenty-first century, especially since the World Wide Web, social networking and online gaming have become part of the fabric of our everyday lives. Though, in a more recent interview Gibson points out that 'cyberspace is so commonplace now it sounds… ridiculous to speak of cyberspace as being somewhere else' (Gibson in Wallace Wells, 2011). So, from the vantage point of the second decade of the twenty-first century, cyberspace appears to have lost some of its initial cachet as an exotic realm which is set apart from our everyday lives and has been replaced by other terms such as 'New Media', which focus on such developments as Web 2.0 and social networking. In his review of *Zero History* for *The New Statesman*, Leo Robson (2010) suggests that William Gibson's novels have moved away from the prophetic and fantastical to mundane descriptions of everyday lives. Similarly, Thomas Jones (2011) laments the ways in which the futuristic features of *Neuromancer* such as the Ono-Sendai Cyberspace desk have been replaced by contemporary technological devices such as the iPhone in *Zero History*. The implication here seems to be that Gibson has lost his prophetic edge and has become mired in observations of contemporary culture. Despite these criticisms, in this chapter I will contend that *Zero History* can be interpreted as a critique of the emphasis that is placed upon surface-level appearances and commodities in contemporary culture and how this fractures the relationships between signs, meaning and the material conditions of reality. But as Gibson's treatment of the issues arising from signification is subtle rather than immediately obvious, this may leave his work open to the charge that it focuses on the prosaic or even mundane aspects of contemporary life.

Branding

Throughout Gibson's later novels there are references to brand names that can be regarded as signs within a signifying sequence that operate within the context of globalization, commodification and cultural capital. For instance, in his study of *Pattern Recognition*, Jameson remarks that Gibson's use of brand names amounts to name-dropping and 'is also a matter of knowledge and an encyclopaedic familiarity with the fashions of world space' (2005, 397). A similar charge could also be made in relation to the references to globally recognized brands such as Apple, Google and Twitter in *Zero History*. On this point, James Purdon (2010) criticizes *Zero History* for the attention that is given to Apple products in the novel. However, in an interview with Mark Ward and Jamillah Knowles on BBC Radio 4 in October 2010, Gibson presents an alternative account of the references to Apple products that appear in his novels. According to Gibson, the representation of Apple products is a way of critically emphasizing brand ubiquity in contemporary culture, and fashion and design are primarily a 'Mac universe'.

The world of advertising, design and branding plays a central role in both *Pattern Recognition* and *Zero History*. In *Pattern Recognition*, the protagonist Cayce has a psychological aversion to brands and therefore wears unbranded clothing, whilst in *Zero History* Hollis investigates the origins of an enigmatic and exclusive clothing brand called Gabriel Hounds for Hubertus Bigend, the chief executive of Blue Ant advertising agency. Within the world of fashion design, marketing and advertising, knowledge of what may become the latest trend is of primary importance, for the ability to anticipate the next trend has the potential to be financially lucrative and is a way to stay ahead in a competitive marketplace. For instance, Bigend tells Hollis that advertising involves brand vision transmission, which is a process of creating narratives around products. According to Bigend, brand vision transmission has become standard within the advertising industry. Consequently, Blue Ant needs to do something different to stand out from other agencies. Bigend tells Hollis that Gabriel Hounds represents a new, cutting-edge mode of brand transmission and asks her to investigate this further and report back to him.

Visual culture

The repeated use of brand names and references to visual culture in Gibson's novels also has implications in terms of the ideal reader, since recognizing and understanding these references require cultural competence in such areas as semiotics, post-structuralism, fashion, design and aesthetics. In addition, as Jameson (2005) has argued, the ability to decipher and understand such references relates to cultural capital and class positioning. Notably, references to visual culture are pervasive within Gibson's oeuvre. For instance, in *Neuromancer* there is a reference to a 'dust-covered Kandinsky coffee table' (1986, 46), which refers to the artist Wassily Kandinsky (1866–1944), who was associated with the renowned Bauhaus School of Art and Design in Germany.[1] Meanwhile, in *Zero History*, Bigend wears a suit that is described as International Klein Blue, which refers to the particular hue used by French artist Yves Klein (1928–1962). Similarly, in *Zero History*, when the character Milgrim visits a gift shop he spots some 'souvenir kitsch' (8) that includes dog-headed angels which have humanoid 'Modigliani-slender' bodies (8), which is a reference to the artist Amedeo Modigliani (1884–1920). My point here is that the ideal reader is invited to connect the references in Gibson's novels to their knowledge of visual culture.[2]

1 The Bauhaus School of Art, Architecture and Design was established in Weimar in 1919 and later moved to Dessau, becoming the School of Design. Source: http://www.bauhaus-dessau.de.

2 Upon close inspection, these references to visual culture in Gibson's novels also seem to follow a rule of three, so we have the Kandinsky coffee table, International Klein Blue and Modigliani slender bodies.

Rootlessness and liquidity

Jameson (1991) also provides a useful perspective on the ways in which postmodernism has instigated a breakdown in terms of the relationships between signifiers and what they signify. In particular, Jameson contends that this breakdown of signification has wider implications in terms of our understanding of temporality and subjectivity. Technology has played a part in this process as digital information can be circulated at great speed across different continents, facilitating the movement of capital in a global financial marketplace. Indeed, Jameson claims that postmodernity has arisen from the historical conditions of neo-liberalism and the global capitalism of the late 1970s and early 1980s. Whilst writing in the wake of the new millennium, Zygmunt Bauman (2000) contends that the global exigencies of contemporary capitalism render employment opportunities fickle and unstable. Digital technologies have facilitated the liquid movement of global capital, but this has also displaced traditional communities and the relationships within them. So not only have signs become de-stabilized in contemporary life but for many people the stability of employment or belonging to a particular community or culture has been eroded, a situation which also resonates with the representation of several characters in *Zero History*.

In *Zero History*, Hollis and Milgrim are rootless; they do not have a secure form of employment or long-term place of residence. Hollis is in a financially precarious situation after losing half of her savings in the stock market crash of 2008. She also lost money when the dot-com bubble burst at the start of the new millennium, which resulted in the closure of her record store in Brooklyn. In this way, Hollis's financial situation can be interpreted as pointing to the instability of contemporary capitalism and the ways in which developed economies have shifted away from the material production of manufacturing towards an abstract global banking and finance system that is predicated upon the circulation of signs. Milgrim's situation is equally precarious; he is a former prescription drug addict for whom the past decade is a blur. Due to his addiction and subsequent treatment, for the past ten years he has not been able to access the Internet or actively participate in the consumption of commodities and has zero credit history. There is an intriguing passage in the novel when Hollis watches Milgrim access the Internet. Hollis remarks that 'he sank instantly into whatever it was that he did on the Net, like a stone into water. He was elsewhere, the way people were before their screens, his expression that of someone piloting something, looking into a middle distance that had nothing to do with geography' (178). This passage suggests that Milgrim is fascinated by accessing the Internet because this is a novel experience for him. In addition, the way in which Milgrim swiftly catches up on what he has missed during the past decade gives the impression that learning how to access and navigate the Internet is effortless. It is also interesting that Milgrim's experience is conveyed through a metaphor that resonates with materiality and substance (stone), but also fluidity (water). So in this passage when the stone becomes immersed in water, the material realm is downplayed and the virtual space of the Net seems to take precedence.

Even though Hollis and Milgrim do not have a stable form of employment or home, they are not subject to the vagaries of contemporary capitalism in the same way as migrant, low-skilled workers. Instead, they are well paid, via a retainer, by Bigend. Hollis and Milgrim are able to secure this arrangement because they have specialist skills such as understanding codes and ciphers. Yet when talking about her employment, Hollis tells another designer Meredith, 'look this is just a job for me, one I wish I didn't have. Not even a job. Just Bigend bribing me to do something for him' (229). This implies that whilst Hollis is not subject to the same demands as a low-paid, casual worker, she is still caught up in an unequal power relationship with her employer, who is seeking to exploit her skills, knowledge and experience.

Tracing the connections between different elements in Gibson's novels, it is possible to argue that they raise a series of anxieties about self-referentiality and the de-stabilization of the relationship between signs and reality. Instead of signs referring back to reality, in terms of the material conditions in which they are produced, exchanged and interpreted, they refer to other signs within a self-referential system of signification. As a result, signs become increasingly free floating as meaning glides across different time periods and between different cultures.

Simulacra

In his study of *Pattern Recognition*, Alex Whetmore (2007) discusses a significant passage in relation to signification and simulacra. In this passage, Cayce goes to the Harvey Nichols department store and unexpectedly comes across racks of Tommy Hilfiger clothing. Cayce finds Hilfiger psychologically nauseating because it is a simulacrum of Ralph Lauren, which in turn is a simulacrum of the clothing found in Brooks Brothers' stores. Ralph Lauren clothing consists of variations on classical designs such as the blazer or chinos, or themed clothing such as horse-riding apparel, or polo. In this way, the Ralph Lauren brand takes designs from one particular context, either a historical period or specific sporting activity, and changes them into signs that operate within the realm of fashion. Similarly, Tommy Hilfiger clothing is predominantly based on a limited colour palette of red, white and blue and retro designs that aim to evoke an imaginary and nostalgic American post-war period. Yet these designs also gain meaning in relation to the designs of previous Tommy Hilfiger collections. On a critical note, Cayce remarks, 'There must be some Tommy Hilfiger event horizon, beyond which it is impossible to be more derivative, more removed from source, more devoid of soul' (Gibson, 2003, 18).[3]

Yet the interpretation of Gibson's work could also be placed in a framework of signs that refer back to other signs, operating across different texts. For instance, *Zero History* harks back to the prior reference to Ralph Lauren and Tommy Hilfiger in *Pattern Recognition*. In *Zero History*, when Milgrim visits Galeries Lafayette in Paris he walks through 'a grove of Ralph Lauren, then a thinner one of Hilfiger' (134). The ways in which Ralph Lauren's clothing line forms a grove but Hilfiger's is a thinner collection also seems to underscore the idea that Hilfiger is derivative of the Lauren brand. So does this repetition of brand names in Gibson's later novels such as *Zero History* point to the surface appearance of circulating signs? Or could this repetition of signs be interpreted as a way of reminding us of the ambiguities and slipperiness of signs and meaning? A further point to consider here is that repetition and difference are creative elements with pattern making and the weaving of literary narratives. To explore these issues in more detail, we will move on to consider how *Zero History* can provide critical insight into the ways in which the rapid circulation of signs within contemporary culture destabilizes the relationship between appearance and reality.

The opening passage to *Zero History* raises a number of questions regarding signification, authenticity and meaning, which require close analysis. The novel opens with a scene in London as Hollis seeks a taxi cab. The reference to the taxi cab is significant on a number of levels, in terms of history, culture and

[3] In 2010, the Tommy Hilfiger brand was acquired by Phillips Van Heusen, a clothing conglomerate which also owns brands such as Donna Karan and Calvin Klein.

changing meanings. Hollis is an American visitor and she reminisces that when she first visited London all the taxi cabs were black. From this vantage point, black cabs can be considered as part of the visual lexicon of London, which also includes red phone boxes and buses. But when Hollis's taxi arrives, it is not a black London taxi; it is pearlescent silver. Hollis describes the cab as 'a smoother simulacrum of its black ancestors' (1). This suggests that the silver cab is a sign that gains meaning in relation to another sign, the black cab. The disjuncture between signs and referents is further underscored when Hollis describes the 'faux-leather upholstery' (1) in the interior of the silver cab, which gains meaning through the difference to real leather, which also raises questions about authenticity.

When the cab arrives, Hollis notices that the exterior is decorated with advertisements for banking services and some sort of business software. This reference to financial services resonates with the setting of the novel in the aftermath of the global economic crisis of 2008. In this way, the opening scene of the novel invites us to think about the relationships between signs, meaning and referents in more detail. Indeed, the black taxis which are associated with London have now become a brand through the establishment of the London Taxi Company, which sells cabs across the world. Yet whilst the London Taxi Company manufactures cabs in the United Kingdom, it is owned by a Chinese firm, Zhejiang Geely Holding Group. In this way the reference to the black cab not only points to a shifting relationship between signs and meanings but also alludes to wider social, cultural and economic changes such as globalization and branding.

In *Zero History*, London is a city that is characterized by a series of meandering, idiosyncratic streets in which different architectural styles are placed side by side. In this way, the meanings associated with the city skim across these different architectural styles from different historical moments. Milgrim states that London had 'perhaps shuffled since the last time he'd been here. The pixels that comprized it were familiar, but it remained only provisionally mapped, a protean territory' (37). What is interesting about Milgrim's remarks is the way in which the materiality of urban space is associated with an image formed by pixels, the flickering signifiers of the computer screen. Yet the urban space of London is only provisionally mapped, which suggests that the totality and the complexity of the city exceed the logic and standardization that can be found in computer programming and image capture.

Whilst in London, Hollis stays in an exclusive private club/hotel called the Cabinet, which has a series of bespoke guest rooms. On her way to her room

Hollis walks down the hallway which is decorated with a series of landscape watercolour paintings. Each painting represents a folly, a building for decoration rather than utility. What is interesting is that Hollis notices that the same folly is used throughout the watercolour collection, even though they represent different geographical locations. So not only is a folly a sort of fake building (e.g. a folly may be built to look like a set of architectural remains), but the way in which the same folly is used in different scenes also unsettles the relationships between signs and the existence of a referent beyond the symbolic system of signification.

After staying at the Cabinet for some time, Hollis realizes that 'one of the landscapes now contained two follies, identical, one further back, on a distant hillside. Surely it had always been there, the second folly unnoticed' (317). This landscape image with identical follies further unsettles the relationships between the scene depicted and the representation of a particular place. The collection of watercolour landscapes no longer represents a particular place; they are now part of the generic images that decorate hotel corridors. In this context the images function as a decorative background rather than providing a prolonged viewing or an intense aesthetic experience. In other words, these images have become like background wallpaper for guests. In a similar way, it could be contended that simulacra permeate contemporary culture to such a degree that we barely notice that the relationships between signs and referents have become destabilized. It is also noteworthy that something about these watercolour images has seeped through into Hollis's consciousness; she becomes aware of the image with two identical follies and draws our attention to it.

Hollis's room in the Cabinet features a strange assemblage of eclectic objects from different historical periods and styles. For instance, there is a bird cage in the room which is suspended from the ceiling and filled with books by different authors and from different eras.[4] Inside the bird cage these books take on a decorative quality which is very different to placing them in an orderly, logical sequence such as the Dewey Decimal System that is used in libraries.[5] In this way the books in the bird cage seem to escape the precision and logic of informatics and take on a more variegated and complex set of meanings. Taken overall,

4 The books in the bird cage in Hollis's room at the Cabinet include *English Eccentrics* by Dame Edith Sitwell and *Rogue Male* by Geoffrey Household.

5 The Dewey Decimal System is named after Melvil Dewey (1851–1931). Source: *The Concise Oxford Companion to English Literature*, Third Edition, edited by Drabble, M., Stringer, J. and Hahn, D. 2007.

I would contend that this passage in the novel points to the ways in which contemporary life is enmeshed within a complex system of signs. Moreover, what seems to emerge from this whirligig of signs is an emphasis on surface appearance: faux-leather, advertisements, follies and interior decoration. Yet, there are also hints that the material conditions in which the signs are produced, circulated and interpreted are not entirely occluded by surface-level appearance. For as the novel progresses, it starts to open out into a more extended and troubling account of the historical, social, cultural and economic impact of the global circulation of signs.

Vintage

As part of her investigation into Gabriel Hounds, Hollis visits a vintage fair of 1980s clothing in Paris with Milgrim. Yet there is a disjuncture between the clothing at the vintage fair, which can be thought of as a series of signs and the ways in which Hollis remembers this time period. Hollis remarks that 'she always found it peculiar to encounter a time that she had actually lived through rendered as a period' (102). These remarks highlight the difference between lived experience, memory and the representational process of history whereby time is ordered into sequences such as periods and decades. On display at the fair is a photograph of Hollis by the famous Dutch photographer Anton Corbjin, which points to a disjuncture between signs and meaning.[6] The image was taken in 1996 when Hollis was in the band 'The Curfew', so it is out of synch with the 1980s theme of the vintage fair. Moreover, Milgrim states that the poster is 'an amateurish reproduction, oversized and lacking in detail' (111). So although a poster could be reproduced many times over, there are differences between an authentic official reproduction and an amateur one. Furthermore, the image was taken at a time when Hollis was famous and her image had cultural and economic currency. Indeed, we could surmise that in the 1990s there were fans of the Curfew that would want to own a poster-size image of her. But when Hollis visits the vintage fair, she is no longer a highly visible figure within popular music. The moment at which Hollis was famous has passed; her image no longer has the same value within the signification system of fashion and music.

[6] Anton Corbjin is renowned for his photographic images of rock stars and videos for bands such as *Depeche Mode* and *U2*.

Hollis has gone to the vintage fair to meet Meredith, who may know something about Gabriel Hounds. Meredith, an Australian, has flown in to Paris to sell clothing at the fair and manages to sell her entire stock of vintage Chanel clothing to a Japanese buyer. In this way, Meredith's activities point to the larger context of recycling signs of the past, such as 1980s vintage clothing and globalization. Indeed, Meredith is an Australian who is selling a clothing line produced by French designer Coco Chanel (1883–1971) to a Japanese collector. The global flow of capital is also underscored when Meredith checks her iPhone to ensure that the financial transaction for the sale has gone into her bank account. In other words, the payment for a vintage line of clothing takes place within the context of digital monetary flows across continents.

Through networking with various contacts within the fashion and music industry, Hollis finally locates the designer of Gabriel Hounds. Yet the identity of the designer remains somewhat mysterious, since she is referred to simply as 'the woman' (335). However, there is a further intertextual dimension to the identity of the designer. From the information the designer gives to Hollis about her background, it is possible for those readers who are familiar with Gibson's oeuvre to deduce that she is Cayce from *Pattern Recognition*. For instance, the designer tells Hollis that she has a psychological aversion to brand names, 'logos, in particular. Corporate mascot figures' (335). When Hollis first sees the Gabriel Hounds designer she is 'seated on one of two old, paint-scarred, mismatched wooden kitchen chairs, stroking the screen of an iPhone' (334), which is a strange juxtaposition of signs and meanings. What I find noteworthy is that there are two chairs but they are not identical; they are mismatched. In this way these chairs contrast with the identical image of the follies in the watercolour landscapes in Hollis's hotel, the Cabinet. Furthermore, the paint scarring on the chairs is a trace of their past use, so these objects have historical and functional resonance. However, the historical materiality of the mismatched chairs is juxtaposed with the reference to the iPhone as a technological device. The designer tells Hollis that her iPhone is an object that enables her to keep in touch with her children, across time zones. So although the iPhone is a material object in the designer's studio, crucially it is a communication device which extends her reach across time and space.

A sense of history and remembrance features strongly in the dialogue that takes place between Hollis and the designer. The designer tells Hollis that she had lived in Chicago and it was there that she 'discovered the ruins of American manufacturing' (336). These remarks point to wider concerns about the shifts that have taken place from the economic powerhouse of heavy manufacturing industries such as coal or ship building in the United States and other countries such as the United Kingdom to an economy based on leisure and consumption. It is worth mentioning here that in contemporary culture consumption is increasingly centred upon brand names. The designer goes on to say that she wanted to make good quality clothing and 'started bumping into people who remembered how to make things' (336). So here we find an emphasis on material forms of production rather than just an emphasis on branded imagery. However, the designer goes on to say that she discovered the Japanese approach to secret brands that operate according to 'parallel microeconomies, where knowledge is more congruent than wealth' (337). The comments made by the designer suggest that instead of buying clothing on the basis of how well it is made, consumer choices are motivated by knowledge of the meanings that accrue around brand names and how they operate within a signifying system. So consumers choose a particular branded product because it is a way of displaying and signalling their place and position in society.

In the context of the novel, Gabriel Hounds is made via small-scale artisan forms of production, so the signs associated with the brand become exclusive. Further limitations are placed on the proliferation of the signs and meanings associated with the brand as the designer tells Cayce that there is no advertising, press releases or fashion shows for Gabriel Hounds. Therefore, recognizing that Gabriel Hounds is a brand that operates in a different way to other brands requires cultural capital such as the knowledge of codes within fashion, advertising and design. Meredith explains the symbolic aspects of Gabriel Hounds clothing to Hollis, stating that it is 'about atemporality. About opting out of the industrialisation of novelty. It's about a deeper code' (116). Yet even atemporarilty is a code within a signifying system; the point is that Gabriel Hounds has a different value to other signs within the signifying system of fashion. Gabriel Hounds is set apart from the fast turnover of designs that are used within mass produced clothing. Instead, Gabriel Hounds is about producing well-made, high-quality clothing that will stand the test of time. Indeed, the representation of Gabriel Hounds clothing in the novel evokes a sense of artisan production through the use of quality materials and small quantities of finished pieces.

A further reference to clothing that disrupts temporality and meaning is made in the novel by reference to an intriguing device called the sigil. Specifically, the sigil is a talisman that generates a sign of forgetting. Hollis's boyfriend Garreth uses a secretive computer code referred to as 'grailware'

(301) to access darknets that are 'private internets, unlicensed and unpoliced' (301). Through accessing his contacts on darknets, Garreth acquires information about 'the sigil of forgetting' (301). He subsequently receives the sigil in the form of a technologically enhanced T-shirt which de-stabilizes the computer code of surveillance cameras. The camera sees the T-shirt and the person wearing it, but instantly forgets them. So the wearer of the T-shirt has zero history as far as the surveillance cameras are concerned. In this way, the concept of the sigil of forgetting also emphasizes the main thematic of the novel in which the relationships between temporality, signs and referents are destabilized.

Incompleteness

The difference between the rationality and order of computer code and the creative turning of linguistic meaning is foregrounded as the narrative of *Zero History* unfolds. For it becomes apparent that Bigend has a much larger agenda than acquiring information about the Gabriel Hounds fashion brand. The mission to find the Gabriel Hounds designer is a McGuffin, a device which enables the characters to come together and to move the narrative forward. In an intriguing passage Milgrim talks about Bigend's larger agenda, stating that it concerns the order flow of the stock market, in other words, knowledge of how the market will operate at any one time. Milgrim goes on to say that all this information exists 'but there's no aggregator' (177). So 'if someone were able to aggregate that [information], the market would cease to be real' (177). Through Milgrim's comments readers are invited to think about the stock market as a dynamic moving picture whereby information is continuously shifting. His comments also suggest that the operation of the stock market involves a fine balance between risk, uncertainty and predictability. Indeed, Milgrim remarks that 'the market is the inability to aggregate the flow at any given moment' (177).

It is also interesting to consider the emphasis on codes in *Zero History* in relation to the term 'Big Data', which has come into recent usage to refer to the process whereby masses of data are brought together from a variety of data sets. Big Data can be used in a variety of ways; for instance, these techniques can be used to help predict disasters and alleviate the destruction they may cause. Indeed, Big Data techniques were used to bring together data sets from satellites, traffic routes and meteorological organizations in relation to Hurricane Sandy, which hit the United States in October 2012. Big Data rests upon the use of mathematical models, a process which raises some interesting debates about human subjectivity, interpretation and decision making. For instance, Baudrillard (1983) considers mathematical modelling in relation to our understanding of reality, claiming: 'The real is produced from miniaturised units, from matrices, memory banks and command models – and with these it can be reproduced an indefinite number of times' (1983, 3). The model 'no longer has to be rational, since it is no longer measured against some ideal or negative instance. It is nothing more than operational' (1983, 3). A major aspect of Baudrillard's work is concern for the ways in which signs within the model generate a sense of reality, rather than referring to an external ontological real. Crucially, Baudrillard contends that these models are regarded as objective and that there is often confusion between 'the fact with its model' (1983, 32). Whilst human interpretation and decision making can be fallible, Baudrillard's comments raise concerns about the problems that could arise when we become so enamoured with these models that we start to cede human control to them. Taking these points into consideration, the ability to bring together different sources of data and find patterns and meaning within them is becoming significant and has a number of practical applications. At the same time, however, we need to be mindful of the differences between mathematical models and reality. Even though the mathematical models of Big Data can yield important information, reality can never be entirely captured and modelled through the accretion of data.

Human subjectivity also remains a crucial aspect to the modelling process because it is important to have people who can construct, utilize and interpret Big Data models effectively. Turning back to Gibson's oeuvre, what we find is that the characters in his novels are often adept at spotting patterns and deciphering codes of various kinds. In *Neuromancer*, for example, the protagonist Case is adept at hacking into corporate computer systems. In *Idoru* (1996) the character Laney is represented as a cybernetic waterwitch who has an intuitive way of understanding patterns of information. And in *Zero History*, Bigend recruits Bobby Chombo, a talented hacker, to find out the order flow of the stock market.[7] Chombo successfully unlocks the order flow of the stock market, 17 minutes into the future. Bigend's quest to find the order flow also

[7] Bobby Chombo is another intertextual character who appears in both *Spook Country* and *Zero History*.

points to the ways in which the stock market has become increasingly abstract and is underpinned by complex mathematical codes.[8] In addition, financial transactions are increasingly represented as signs that zap across continents almost instantaneously.

In *The Physics of Wall Street*, James Owen Weatherall (2013) provides an overview of the ways in which physicists and mathematicians are employed by Wall Street companies to devise complex and abstract financial models consisting of quants and derivatives. What is noteworthy is that Weatherall refers to the scientist Jim Simons, who created a mathematical model based on string theory to create an investment portfolio. According to Weatherall, Simons used his knowledge of mathematics and physics to predict the unpredictable, making a fortune on the stock market, despite the global financial crisis of 2008. In this way, Bigend's fictional quest to control the stock market in *Zero History* seems to evoke the financial activities of Jim Simons and contemporary concerns about the increasing abstraction of global finance.

Gibson's novel also creates parallels between an abstract, speculative stock market and cyberspace. Bigend ends up owning a large proportion of Iceland after he persuaded 'young Icelandic fiscal cowboys on the idea of internet banking' (403). The reference to cowboys is particularly evocative of Case in *Neuromancer*, who is referred to as a cyberspace cowboy. Therefore, the use of the term 'fiscal cowboy' suggests that the global economic system exists in the form of cyberspace, a computer-generated code world. Yet as we have seen, when Iceland moved from an economy that was primarily based on local forms of production such as fishing to global finance, the results were disastrous as the country's banking system collapsed (Wade and Sigurgeirsdottir, 2011). However, the winners of this new abstract economy seem to be those who can draw upon intellectual capital in the form of specialized skills in physics and mathematical modelling, since this can be used to optimize the global flow of finance and yield financially beneficial results.

The characters in Gibson's novels are often recruited on the basis of their specialist knowledge and ingenuity since this enables them to decipher codes which others find baffling. In this way, human subjectivity is foregrounded as a

[8] Codes are a persistent theme within Gibson's work. In *Neuromancer*, for example, there are codes that generate security systems known as Intrusion Countermeasures Electronics (ICE). Meanwhile, in *Spook Country*, Volapük is a code which is used for the purposes of espionage. Volapük is a constructed, artificial language that was produced in 1879 by Johann M Schleyer. Source: *Oxford English Dictionary*, third edition, edited by Angus Stevenson, 2010.

way of interpreting codes, of making sense of the complex symbolic signs that are operating in contemporary culture. So although Gibson's work is often interpreted as highlighting computer technology and artificial intelligence, a different viewpoint can be taken here in which human subjectivity is foregrounded. Arguably, this foregrounding of human ingenuity and interpretation in Gibson's novels can help us make sense of the ways in which technologies intersect our lives and consider how this might impact upon subjectivity. In addition, Katherine N. Hayles (2005) reminds us that there are important differences between literature and computation, since the former operates through figuration and tropes and generates a plurality of meanings. In contrast, Big Data operates according to algorithmic functional parameters in which ambiguity must be eradicated. Therefore, Gibson's novels can be taken as a reminder that creativity cannot be mapped and analysed in the same way as Big Data.

Zero History ends on an enigmatic note in which Hollis is dreaming. The content of Hollis's dream recombines some of the elements of the novel, but they are twisted and turned to create new meaning. For instance, at the end of the novel, Hollis finds herself in a room on an unknown floor which is 'faded antediluvian' where there are 'walls hung with madly varied landscapes' (404). This seems to be a way of recombining the theme of the novel and also challenging the emphasis on simulacra in contemporary culture. For the use of the term 'antediluvian' could be interpreted as something antiquated, a time before a deluge of signs permeated everyday life. In addition, the room is faded, which suggests that it has changed through the course of time. Hollis is no longer staying at the Cabinet and viewing the images on the corridors which represent the same folly in different landscapes. Instead, Hollis is dreaming about viewing images that represent a variety of landscapes. Finally, the reference to madly varied landscapes suggests a form of creativity that can take us beyond the rational logic of computer code where signs and meanings must exactly correspond.

To recap, in this chapter we have seen that in *Neuromancer* Gibson set out a futuristic vision of cyberspace as the site of power struggles between corporations and sub-cultural groups such as hackers. When *Neuromancer* was first published in the early 1980s, the Internet and World Wide Web were not commercially available and the idea of accessing a huge network of information and becoming immersed within computer-generated spaces was an imaginative and exciting new idea. Although the vision of complete sensory immersion in virtual reality has not come to fruition, I have argued that *Zero History* points to the ways in which virtuality has slipped into the interstices of our daily lives. Drawing upon the work of Jameson and Baudrillard, this chapter has indicated that Gibson's novel concerns the proliferation and speed at which signs are circulated and how this impacts upon subjectivity, agency and temporality. This is exemplified by the ways in which finance is increasingly represented as digital signs that can move across continents almost instantaneously and the impact this has in terms of macro and micro economics. For instance, the global circulation of finance may impact upon the sort of work we do, where we work and where we live. In this chapter we have seen that Hollis, the protagonist in *Zero History*, works on a retainer for Hubertus Bigend, the chief executive of the London-based advertising agency Blue Ant. Although the novel mentions that Hollis has lost money in the financial crash, she is not in a financially precarious position. Instead, she appears to be well paid by Bigend, since she stays in an up-scale hotel, eats in fine restaurants and networks with fashion designers and famous musicians. In this regard, Hollis is in a rather privileged position compared to unskilled, migrant workers or those whose skills have been displaced by the changes that have taken place within the global capitalist system and the shift in emphasis from manufacturing to financial services.

As discussed, Gibson's novel is about culture, the production and circulation of meaning within the creative industries of fashion and design. Throughout the book there is continual reference to what the characters are wearing rather than their emotional state of mind. In this way, the novel seems to encourage the reader to imagine what the characters look like rather than gaining insight into their thoughts or feelings. Taking this point into consideration it could be charged that Gibson's characters lack depth but perhaps that is the point, since the novel itself concerns the ways in which the meaning of signs have become increasingly detached from material conditions.

At another level, however, the novel does call our attention to how things are made. But *Zero History* does not attempt to expose the fraught social relations and inequalities within the global, mass-produced fashion industry in the same way as Naomi Klein's non-fiction work *No Logo* (2000) or a social realist novel. Instead, Gibson's novel gestures towards the differences between the rapid turnover of designs within mass-produced clothing and the timelessness of high-quality clothing and artisan production. Even so, Gibson's novel does not appear to be a call to overthrow industrial production and return to hand crafted, small-scale modes of production. Indeed what we find in the novel is that artisan production rests alongside global communications and finance. For example, we have seen that when Meredith sells couture clothing, she uses her iPhone to check that an electronic payment has reached her bank account. And the Gabriel Hounds designer works in an artisan-like studio in London yet also uses an iPhone to keep in touch with her family in the United States.

This chapter has also indicated that there are strong inter-textual dimension to *Zero History*, for instance, we have seen that there are similarities between Hollis and Cayce in *Pattern Recognition* – both Hollis and Cayce are recruited by Bigend to find out the source of a mysterious form of cultural production. Cayce is recruited to track down the producer of a strange collection of film footage that is gaining a cult following through circulation on the Internet and Hollis is recruited to track down the designer of Gabriel Hounds clothing. Taking a wider perspective, Hollis and Cayce are characters that are also reminiscent of those found in other Gibson novels such as Laney in *Idoru*, who has the ability to sift through information and find valuable nuggets of information and patterns within data, and even Case in *Neuromancer*, who can circumnavigate the patterns of code that generate the firewalls of corporate on-line systems. Taking these intertextual references into consideration, it seems that throughout Gibson's oeuvre there are concerns about being in the know, of having specialist knowledge and cultural capital. This emphasis of being in the know also has implications in terms of the readership of Gibson's novel. Readers who are familiar with his work will be able to recognize these inter-textual references and connections and in this way they are drawing upon their cultural capital. Yet this is not to say that Gibson's work is simply repetitive, instead he twists and turns the meanings that are associated with the themes and characters across different novels. In this way his use of language contrasts with computer code whereby signs operate logically and rationally and have fixed meanings. So although Gibson's work is often associated with a preoccupation with computer technology, his use of evocative, metaphorical language emphasizes the ambiguity between signs and meaning.

By the end of the novel Bigend achieves his quest to obtain knowledge as a means of increasing his wealth and power. In this way, abstract mathematical models are used to fulfil human desires such as increased personal wealth and power. Though, what is interesting about *Zero History* is that Hollis meets the designer of the Gabriel Hounds brand but she does not reveal this information to Bigend. Furthermore, Hollis acquires information about the designer through networking with other people, rather than sifting through data online. From this vantage point, it seems that for Hollis the bonds of human connection go deeper than Bigend's commercial imperatives.

As the novel finishes, readers are left with an enigmatic ending that invites us to consider the meanings of Hollis's dream. So although the novel ends in the virtual world of the dream, it also invites us to contemplate the ways in which our everyday lives are becoming a mash-up of different elements from different cultures and historical periods. At one level, Gibson's novel seems to revel in ambiguity as illustrated by the madly varied landscapes of Hollis's dream. In this way, *Zero History* can be interpreted as a way of foregrounding creativity and human agency instead of the logic and precision of digital code. Yet the novel also points to the ways in which digital codes have slipped into the interstices of our daily lives, have the potential to disrupt the relationships between signs and meanings and impact upon the economic realm. But in the end, it would seem that Gibson's novel points to the ways in which shifts towards an abstract virtual economy of signs favour those who have high levels of skills in constructing and deciphering codes (computer programming, or cultural codes of fashion and advertising).

In these closing remarks it is interesting to consider the ways in which the shift towards the rapid and global circulation of signs is also impacting the publishing industry. For instance, sales of the paperback or hardback novel are changing due to the introduction of e-books and the circumvention of publishing copyrighted materials on the World Wide Web. In this regard, Gibson is a well-known author with a substantial, critical and popular track record of success. In a way, Gibson is an author who has become a brand, signifying a certain kind of writing that will appeal to certain readers. From this perspective, Gibson's novels are caught up in the profusion of signs that circulate amongst publishers, marketing and advertising companies and readers.

In conclusion, this chapter also works at the level of signs, producing linguistic signifiers that point to other texts (Gibson's novels) to stimulate debates about signification and meaning in contemporary culture. But what this chapter set out to do was to connect the idea of rapidly circulating free-floating signifiers to actual material conditions, such as globalization and the shift from manufacturing to speculative finance. In the end, *Zero History* does have a history; it has been produced in particular social, cultural and economic circumstances. But one criticism here is that what are less visible in the novel are those members of society who do not have the economic means to purchase exclusive artisan clothing or iPhones or the ability to turn the operational parameters of a speculate stock market into massive personal gain.

Conclusion

This book has indicated that attempts to explore immersive virtual realities have a complex historical trajectory. Indeed, throughout history and in different cultural contexts attempts have been made to explore the virtual and the real through incantations, hallucinogenic plants and various forms of visual representation such as Trompe L'Oeil, the Panorama, 3D cinema and computer-generated spaces. To begin with, this book presented a brief historical review of the development of virtual reality in order to contextualize the ways in which this technology and the ideas that were associated with it were particularly prominent during the 1980s and 1990s. As discussed, in the 1980s popular representations of virtual reality were connected with the burgeoning computer games market and found expression in contemporary film, science fiction literature and other media. Then, by the 1990s virtual reality was entwined with technological, political and economic debates about the Information Superhighway.

This book has paid particular attention to the ways in which representations of virtual reality in contemporary media popularized the idea of using technology to transcend the limitations of the physical body. However, upon close inspection representations of virtual reality in a range of popular films were regarded as multi-layered, producing multiple levels of interpretation. In fact, major studios produce films that feature both critical and celebratory representations of virtual reality. Furthermore, as we have seen, even within one specific film, such as *Source Code* or *Avatar*, there may be multiple and contradictory representations of virtual reality technology.

In the second decade of the twenty-first century it could be contended that William Gibson's concept of cyberspace as a new computer-generated space has become ubiquitous. From this perspective, it is important to consider whether there may be generational differences between those who remember and lived through the pre-Internet age and the generation born in the 1990s and onwards, in the era of digital technologies. Indeed, in the book we considered Marc Prensky's (2001) distinction between those born before the Internet age who have made forays into technologically mediated spaces (digital immigrants) and the Internet generation (digital natives).

Today, the abstract immaterial realm of mathematics and computer code provides a global information infrastructure that impacts our lives in many different ways. For example, we have considered the ways in which the global financial crisis of 2008 relates to abstract mathematical processing and flickering virtual images that represent financial values. But we also considered the material impact arising from the operations of the crisis that occurred in global financial markets in terms of job losses and the foreclosure of homes.

This book has also considered the ways in which the idea of transcending the limitations of the physical realm and becoming immersed in virtual worlds contains an element of wish-fulfilment. In this regard, technology appears to offer the possibility of escaping from our world and the problems we face into a virtual world in which time, space and embodiment are no longer constraining. In this regard we have also explored the ways in which computer scientist Ray Kurzweil even positions mortality as a technological problem which can be overcome by living a post-embodied life in virtual reality.

Taking these points into consideration it may seem churlish to criticize the technological wonders of virtual reality, especially if technology can be used to augment our physical bodies, our mental capabilities and even transcend mortality. Furthermore, perhaps it is sentimental to cherish embodied forms of social interaction. In fact, digital natives may not be so concerned about the distinction between the virtual and the real, or prioritize embodied social interaction over technologically mediated forms of communication. Yet, in this book I suggest that it is not simply sentimental or anachronistic to raise problematic questions about the relationships between technology and embodiment, for we remain embodied and embedded within the world around us, even if that world is intertwined with digital technology.

This book has also shown that there can be gaps between our online virtual selves and the actual material conditions that impact our everyday lives. In this regard there may also be important differences between our idealized, constructed virtual selves and how we perform different aspects of our identity online via social networks or MMORPGs, and our face-to-face interactions with others. So, whilst we may be able to create and explore other identities through avatars or online profiles, class, race, ethnicity, gender, age and disability continue to be important markers that shape our identities and interactions with others. During the discussion of the film *Catfish*, for example, we saw the unravelling that can occur between idealized online forms of identity and face-to-face interaction.

Technologies such as virtual reality do not simply determine social or cultural attitudes towards embodiment, or face-to-face interaction. Rather, as this book has indicated, technological development is interwoven with social and cultural practices and other factors such as economic or political change. Therefore, some aspects of technological development are unforeseen. For example, whilst the possibilities of immersive virtual worlds were particularly prominent in contemporary media during the 1990s, the massive growth and popularity of online social networking were less noteworthy during this period. In the 1990s, contemporary media was replete with visions of incredibly immersive virtual experiences in films such as *Tron*, *The Lawnmower Man* and the early novels of William Gibson. But in the twenty-first century, instead of finding ourselves within these immersive virtual worlds, we are now simultaneously engaging in online and offline activities and environments through mobile computing and smart phone technology. Even so, representations of virtual reality in contemporary media can be regarded as offering an important site in which to consider and debate how technology may shift our sense of the virtual, the real and embodiment. On this note, we have considered how *Strange Days* offers a vision of immersive virtual technology and how this might be utilized as a way of disengaging from social interaction and escaping the real social and economic problems that exist in contemporary urban environments. But instead of the playback technology which is represented in *Strange Days*, in the twenty-first century mobile technologies such as the Apple iPod are a popular way of personalizing our experiences of social spaces, avoiding social contact and becoming immersed within our own private auditory bubbles (Bull, 2004).

Whilst our sense of embodiment or embodied habits may change due to our interaction with technology, there are still some aspects of human relationships that cannot be fully technologically replicated, such as the sense of smell or touch. So even in the second decade of the twenty-first century human relationships are still connected to embodied bio-chemical markers. On this basis there are still important ontological and epistemological distinctions to be made between the virtual and the real, especially in relation to embodiment, since real bodies feel pain, suffer and die.

The critical analysis of contemporary representations of virtual reality presented in this book has been guided by theory, engagement with academic and popular debates. At the same time, the discussion is set in a backdrop of technological complexity and uncertainty. In this regard the book presents an open and honest conversation that attempts to contribute to debates about technological developments and our understanding of the virtual and the real. Although the trajectory of technological development, social and cultural change is uncertain, it is still important to explore the ways in which virtual reality as a philosophical concept and a particular technological form is represented, interpreted and understood.

References

Abbott, Alison. 'Art That Draws You In.' *Nature* 427 (2004): 17.

Adams, Fredrick. 'The Informational Turn in Philosophy.' *Minds and Machines* 13 (2003): 471–501.

Adams, Guy. 'Blue Aliens Help Hollywood to Its Best Weekend at the Box Office.' *The Independent*, Tuesday 29th December (2009).

Allen, Michael. 'Digital Cinema: Virtual Screens – Case Study: Star Wars Episode II: Attack of the Clones' In *Digital Cultures Understanding New Media*, ed. Glen Creeber and Martin Royston, 61–75. Maidenhead: Open University Press, 2009.

Amos, Martyn. *Genesis Machines: The New Science of Biocomputation*. London: Atlantic, 2006.

Anderson, John. 'Still Creating Otherwordly Adventures.' *New York Times*, 1st February, available from: http://www.nytimes.com [accessed 2nd April 2013].

Anderson, John. 'Still Creating Othewordly Adventures.' *New York Times*, 1st February (2012), available from http://www.nytimes.com/2012/02/05/movies/awardsseason/ [accessed 20th August 2013].

Andrews, Lori. *I Know Who You Are and I Saw What You Did – Social Networks and the Death of Privacy*. New York: Free Press, 2013.

Ascott, Ray. ed. *Reframing Consciousness*. Exeter: Intellect, 1999.

Bachelard, Gaston. *The Poetics of Space*. Boston: Beacon Press, 1994.

Bailey, Nick. 'Technology at the Service of Art.' *Creatie* 9, no. 3, June (2012): 44–47.

Balsamo, Anne. 'Signal to Noise: On the Meaning of Cyberpunk Subculture.' In *Communication in the Age of Virtual Reality*, ed. Frank Biocca and Mark Levy, 347–368. New Jersey: Lawrence Erlbaum Associates, 1995.

Bannerman, Lucy. '3D Blockbuster's Spectacular Success Leads Way to Record Take at Christmas Box Office.' *The Times* (Scotland), Tuesday 29th December (2009).

Barnard, Malcolm. *Approaches to Understanding Visual Culture*. Basingstoke: Palgrave, 2000.

Barrionuevo, Alexi. 'Brazil, After a Long Battle Approves an Amazon Dam.' *New York Times*, 2011, 1st June available from http://www.nytimes.com/2011/06/02/world/americas/02brazil.html [accessed 20th August 2013].

Bartlett, Mark. 'Going (Digitally) Native.' *Animation: An Interdisciplinary Journal* 7, no. 3 (2012): 287–307.

Baudrillard, Jean. *Simulations*. New York: Semiotext(e), 1983.

Baudrillard, Jean. *The Ecstasy of Communication*. Translated by Bernard and Caroline Schutze. New York: Semiotext(e), 1988.

Baudrillard, Jean. *Fatal Strategies*. Translated by Philip Beitchman and W.G.J. Niesluchowski. New York and London: Semiotext(e) and Pluto, 1990.

Baudrillard, Jean. *Simulacra and Simulations*. Translated by Sheila Faria Glaser. Michigan: University of Michigan Press, 1994.

Baudrillard, Jean. *The Gulf War Did Not Take Place*. Translated by Paul Patton. Sydney: Power Publications, 1995.

Baudrillard, Jean. *The Perfect Crime*. Translated by Chris Turner. London: Verso, 1996.

Baudrillard, Jean. *The Consumer Society*. London, Thousand Oaks, New Delhi: Sage Publications, 1998.

Baudrillard, Jean. *The Vital Illusion*. New York and Chichester West Sussex: Columbia University Press, 2000.

Baudrillard, Jean. *Screened Out*. Translated by Chris Turner. London and New York: Verso, 2002.

Bauman, Zygmunt. *Liquid Modernity*. Cambridge: Polity, 2000.

Baxter, Stephen. *The Science of Avatar*. London: Victor Gollancz, 2012.

Baym, Nancy. K. *Personal Connections in the Digital Age*. Cambridge: Polity, 2010.

Bell, Daniel. *An Introduction to Cybercultures*. London and New York: Routledge, 2001.

Benjamin, Walter. *Illuminations*. London: Fontana, 1973.

Berger, John. *Ways of Seeing*. London: British Broadcasting Corporation, 1972.

Berrettini, Mark. 'Can We All Get Along? Social Difference, the Future and Strange Days.' *Camera Obscura* 17, no. 2 (2002): 155–189.

Bethke, Bruce. 'Cyberpunk.' *Amazing Science Fiction Stories* 57, no. 4, November (1983), available from: www.infinityplus.co.uk/stories/cpunk.htm [accessed 20th August 2013].

Bordwell, David. *The Way Hollywood Tells It: Story and Style in Modern Movies*. Berkley: University of California Press, 2006.

Botha, Catherine Frances. 'Heidegger, Technology and Ecology.' *South African Journal of Philosophy* 22, no. 2 (2003): 157–172.

Boulter, Jay David. 'Remediation and the Desire for Immediacy.' *Convergence* 6, no. 1 (2000): 62–71.

Boulter, Jonathan. Stuart. 'Partial Glimpses of the Infinite: Borges and the Simulacrum.' *Hispanic Review* 69, no. 3 (2001): 355–377.

Bourdieu, Pierre. *Distinction: A Social Critique of the Judgement of Taste*. Translated by Richard Nice. London: Routledge, 1986.

Bourdieu, Pierre. *The Field of Cultural Production Essays on Art and Literature*, ed. Randal Johnson. Cambridge: Polity, 1993.

Braidotti, Rosi. *Metamorphoses – Towards a Materialist Theory of Becoming*. Cambridge: Polity, 2002.

Braidotti, Rosi. 'Posthuman, all too Human: Towards a New Process Ontology.' *Theory, Culture and Society* 23, no. 7–8 (2006): 197–208.

Broadhurst Dixon, Joan and Eric Cassidy. eds. *Virtual Futures – Cyberotics, Technology and Post-Human Pragmatism*. London and New York: Routledge, 1998.

Brooke, Heather. *The Revolution Will Be Digitalised – Dispatches from the Information War*. London: William Heinemann, 2011.

Brouillette, Sarah. 'Corporate Publishing and Canonization: Neuromancer and Science-Fiction Publishing in the 1970s and Early 1980s.' *Book History* 5 (2002): 187–208.

Bryant, Antony. 'Of Mice and Mien or Perhaps Mouses and Mien (Anyway with Apologies to John Steinbeck).' In *Digital and Other Virtualities*, ed. Antony Bryant and Griselda Pollock, 42–62. New York: IB Tauris, 2010.

Buchanan, Ian. *A Dictionary of Critical Theory*. Oxford: Oxford University Press, 2010.

Buick, Joanna. 'Virtual Reality and Art.' In *Computers and Art*, ed. Stuart Mealing, 109–118. Bristol: Intellect, 2002.

Bukatman, Scott. *Terminal Identity: The Virtual Subject in Postmodern Science Fiction*. Durham: Duke University Press, 1993.

Bukatman, Scott. 'Who Programs You? The Science Fiction of the Spectacle.' In *Liquid Metal the Science Fiction Film Reader*, ed. Sean Redmond, 228–238. London and New York: Wallflower Press, 2004.

Bull, Michael. 'No Dead Air! The iPod and the Culture of Mobile Listening.' *Leisure Studies* 24, no. 4 (2004): 343–355.

Burgess, Marilyn. Interview with Char Davies, available from: http://www.sidestreet.org/sitestreet_fall2002 [accessed 5 December 2012].

Burkeman, Oiver. 'Personal Profiles: Jaron Lanier.' *The Guardian*, Manchester (UK), 29th December (2001).

Burnett, Ron. *How Images Think*. Cambridge, MA: MIT Press, 2004.

Buttimer, Anne and David Seamon. eds. *The Human Experience of Space and Place*. London: Croom Helm, 1980.

Caldwell, John Thornton. ed. *Electronic Media and Technoculture*. New Brunswick, NJ: Rutgers University Press, 2000.

Calhoun, Craig. ed. *Dictionary of the Social Sciences*. Oxford: Oxford University Press, 2002.

Cameron, James. *Strange Days*. London: Penguin, 1995.

Cameron, James. *Strange Days*. London: Penguin, 1996.

Camille, Michael. 'Simulacrum.' In *Critical Terms for Art History*, ed. Robert S. Nelson and Richard Shiff, 35–50. Chicago and London: University of Chicago Press, 2003.

Campbell, Joseph. *The Hero with a Thousand Faces*. Princeton, NJ: Princeton University Press, 1973.

Campbell, Joseph. *The Hero with a Thousand Faces*. London: Fontana, 1993.

Carr, Brian. 'Strange Days and the Subject of Mobility.' *Camera Obscura* 17, no. 2 (2002): 191–216.

Carter, Denise. 'Living in Virtual Communities – An Ethnography of Human Relationships in Cyberspace.' *Information, Communication and Society* 8, no. 2 (2005): 148–167.

Castells, Manuel. *The Rise of the Network Society*. Chichester: Wiley-Blackwell, 2010.

Castronova, Edward. *Synthetic Worlds: The Business and Culture of Online Games*. Chicago: University of Chicago Press, 2005.

Castronova, Edward. *Synthetic Worlds*. Chicago: University of Chicago Press, 2006.

Cavallaro, Dani. *Cyberpunk and Cyberculture*. London: The Athlone Press, 2000.

Christie, Ian. "Clash of the Wonderlands 3D Cinema" *Sight and Sound*, 10 February 2012, available from http://old.bfi.org.uk/sightandsound/feature/49789 [accessed 19th August 2013].

Concannon, Kevin. 'The Contemporary Space of the Border: Gloria Anzaldua's *Borderlands* and William Gibson's *Neuromancer*.' *Textual Practice* 12, no. 3 (1998): 429–442.

Corliss, Jonathan 'Introduction: The Social Science Study of Video Games.' *Games and Culture* 6, no. 3 (2011): 3–16.

Corneliussen, Hilda and Jill Walker Rettberg. eds. *Digital Culture, Play and Identity: A World of Warcraft Reader*. Cambridge MA: MIT Press, 2008.

Coyle, Rebecca. 'The Genesis of Virtual Reality.' In *Future Visions, New Technologies of the Screen*, ed. Hayward, Philip and Tana Wollen, 148–166. London: British Film Institute, 1993.

Crang Mike, Philip Crang, & Jon May. ed. *Virtual Geographies*. London and New York: Routledge, 1999.

Crick, Timothy. 'The Game Body: Toward a Phenomenology of Contemporary Video Gaming.' *Games and Culture* 6, no. 3 (2011): 259–269.

Csicsery-Ronay, Istvan. 'The Sentimental Futurist: Cybernetics and Art in William Gibson's Neuromancer.' *Critique* 33, no. 3 (1992): 221–240.

Cubitt, Sean. *Simulation and Social Theory*. London: Sage, 2001.

Damasio, Antonio, R. *The Feeling of What Happens – Body, Emotion and the Making of Consciousness*. London: Vintage, 2000.

Damasio, Antonio R. 'Descartes Error Revisited.' *Journal of the History of the Neurosciences* 10, no. 2 (2001): 192–194.

Damasio, Antonio, R. *Descartes' Error: Emotion, Reason and the Human Brain*. London: Vintage, 2006.

D'Amato, Brian. 'Virtual Kitsch.' *Art Forum* 34, no. 5 (1996): 35–36.

Darley, Andrew. *Visual Digital Culture: Surface Play and Spectacle in New Media Genres*. London: Routledge, 2000.

Davies, Char. 'As if.' *Forbes* Supplement 164, no. 8 (1999): 161.

Davies, Char. 'Landscape, Earth, Body, Being, Space, and Time in the Immersive Virtual Environments Osmose and Ephemere.' In *Women, Art and Technology*, ed. Judy Malloy, 322–337. London: MIT Press, 2003.

Davis, Erik. *Techngnosis – Myth, Magic and Mysticism in the Age of Information*. London: Serpent's Tail, 1999.

Davis, Mike. *City of Quartz*. London: Verso, 1990.

Davies, Nick. *Flat Earth News*. London: Vintage, 2009.

Debord, Guy. *Society of the Spectacle*. New York: Zone, 1994.

De la Peňa, Nonny., Peggy Weil, Joan Llobera, Elias Giannopoulous, Ausias Pomĕs, Bernard Spanlang, et al. 'Immersive Journalism: Immersive Virtual Reality for the First Person Experience.' *Presence-Teleoperators and Virtual Environments* 19, no. 4 (2010): 291–301.

Delaney, Bob. 'New Developments: Virtual Reality'. US Senate Subcommittee Hearing, 1991. Cyberedge Information Services, available from: http://www.cyberedge.com [accessed 19 March 2005].

Deleuze, Gilles. *Différence et Répétition*. Translated by Paul Patton. New York and London: Continuum, 1997.

Descartes, René. *The Philosophical Writings of Descartes*. Translated by John Cottingham, Robert Stoothoff and Dugald Murdoch. Cambridge: Cambridge University Press, 1985.

Deutsch, David. *The Fabric of Reality*. London: Allen Lane, Penguin, 1997.

Dick, Philip.K. *Do Androids Dream of Electric Sheep*. London: Rapp & Whiting, 1969.

Dick, Philip K. *UBIK*. London: Vintage, 1969.

Dietz, Steve. 'Ten Dreams of Technology.' *Leonardo* 35, no. 5 (2002): 509–522.

Doel, Marcus. A. and David Clarke. 'Virtual Worlds-Simulation, suppletion, s(ed)uction and simulacra.' In: *Virtual Geographies*, ed. M. Crang, P. Crang and J. May, 261–283. London and New York: Routledge, 1999.

Doel, Marcus., A Clarke and B David. 'Virtual Worlds-Simulation, Suppletion, S(ed)uction and Simulacra.' In *Virtual Geographies*, ed. Mike Crang, Phil Crang and Jon May, 261–283. London: Routledge, 1999.

Drabble, Margaret., Jenny Stringer and Daniel Hahn. *The Concise Oxford Companion to English Literature*. Third Edition. Oxford: Oxford University Press, 2007.

Dunst, Alexander. 'Late Jameson, or After the Eternity of the Present.' *New Formations* 65 (2008): 105–118.

Dyson, Frances. *Sounding New Media: Immersion and Embodiment in the Arts and Culture*. Berkeley: University of California Press, 2009.

Eagleman, David. *Incognito-The Secret Lives of the Brain*. London: Canongate, 2012.

Ebert, Roger. 'Strange Days', 13 October 1995, available from: http://rogerebert.suntimes.com [accessed 14 November 2012].

Ebert, Roger. 'Review of Catfish.' *Chicago Sun Times*, 22nd September 2010, available from: http://rogerebert.suntimes.com [accessed 30th August 2011].

Eco, Umberto. *Travels in Hyperreality*. Translated William Weave. London: Harvest, 1986.

Egan, Greg. *Permutation City*. London: Orion, 1994.

Elsaesser, Thomas. 'James Cameron's Avatar: Access for All.' *New Review of Film and Television Studies* 9, no. 3 (2011): 247–264.

Fair, Benjamin. 'Stepping Razor in Orbit: Postmodern Identity and Political Alternatives in William Gibson's Neuromancer.' *Critique* 45, no. 2 (2005): 92–103.

Featherstone, Mike and Roger Burrows. eds. *Cyberspace, Cyberbodies, Cyberpunk: Cultures of Technological Embodiment*. London: Sage, 1995.

Foucault, Michel. *The Order of Things: An Archaeology of the Human Sciences*. New York: Vintage, 1994.

Franck, Karen. 'When I Enter Virtual Reality, What Body Will I Leave Behind?' In *Cyber_Reader, Critical Writings for the Digital Era*, ed. Neil Spiller, 238–245. London and New York: Phaidon, 2002.

French, Philip. 'It's a Jungle Out There in Outer Space: At $500m, Avatar Is the Most Expensive Movie Ever. Yet for All Its Brilliant Imagery, Is It Any More than a Smug Sermon?' *Observer* 20 December (2009).

Frenkel, Karen. A. 'A Conversation with Jaron Lanier.' *Interactions* July (1995): 47–72.

Friedman, George. 'The Future of War: A Farewell to Arms?' *Psychology Today* November–December (1997): 62.

Gates, Bill. *The Road Ahead*. London: Penguin, 1996.

Gauntlett, David. *Making Is Connecting-The Social Meaning of Creativity, from DIY and Knitting to YouTube and Web 2.0*. Cambridge: Polity, 2011.

Genosko, Gary. *Baudrillard and Signs – Signification Ablaze*. London and New York: Routledge, 1994.

Gibson, William. *Neuromancer*. London: Grafton, 1984.

Gibson, William. *Count Zero*. New York: Arbor, 1986.

Gibson, William. *Virtual Light*. London: Viking, 1993.

Gibson, William. *Idoru*. London: Viking, 1996.

Gibson, William. *Pattern Recognition*. London: Penguin, 2003.

Gibson, William. *Spook Country*. London: Penguin, 2008.

Gibson, William. *Zero History*. London: Viking-Penguin, 2010.

Giddens, Anthony. *The Third Way: The Renewal of Social Democracy*. Cambridge: Polity Press, 1999.

Goffman, Erving. *The Presentation of Self in Everyday Life*. Garden City, New York: Doubleday, 1959.

Gore, Al. 'US Senate Subcommittee Hearing.' *Cyberedge Journal*, 1991. http://www.cyberedge.com [accessed 19 Mar 2005].

Graham, Beryl. *Serious Games. Exhibition Catalogue*. Newcastle Upon Tyne: Barbican Art Gallery and Tyne and Wear Museums, 1996.

Grau, Oliver. *Virtual Art from Illusion to Immersion*. Cambridge, MA and London: MIT Press, 2003.

Grosz, Elizabeth. *Volatile Bodies: Toward a Corporeal Feminism*. Bloomington: Indiana University Press, 1994.

Guest, Tim. *Second Lives – A Journey through Virtual Worlds*. London: Hutchinson, 2007.

Gunkel, David. 'The Real Problem: Avatars, Metaphysics and Online Social Interaction.' *New Media Society* 12, no. 1 (2010): 127–141.

Hansen, Mark. *New Philosophy for New Media*. Cambridge, MA: MIT Press, 2004.

Haraway, Donna. *Simians, Cyborgs and Women: The Reinvention of Nature*. London: Free Association Press, 1991.
Harkaway, Nick. *The Blind Giant: How to Survive in the Digital Age*. London: John Murray, 2012.
Hayles, Katherine. N. 'Embodied Virtuality.' In *Immersed in Technology, Art and Virtual Environments*, ed. Douglas Macleod and Mary Anne Moser, 1–28. Cambridge: MIT Press, 1996.
Hayles, Katherine, N. *How We Became Posthuman*. Chicago: University of Chicago Press, 1999.
Hayles, Katherine N. *My Mother Was a Computer – Digital Subjects and Literary Texts*. Chicago and London: University of Chicago Press, 2005.
Hayward, Philip and Tana Wollen. eds. *Future Visions, New Technologies of the Screen*. London: British Film Institute, 1993.
Hearn, Alison. 'Structuring Feeling: Web 2.0, Online Ranking and Rating, and the Digital "Reputation" Economy.' *Ephemera* 10, no. 3–4 (2010): 421–438.
Heidegger, Martin. 'The Question Concerning Technology.' In *Basic Writings: From Being and Time(1927) to the Task of Thinking (1964)*, ed. David Farrell Krell, 308–341. New York: Harper Collins, 1993.
Heim, Michael. *The Metaphysics of Virtual Reality*. New York and Oxford: Oxford University Press, 1993.
Heim, Michael. 'The Design of Virtual Reality.' In *Cyberspace, Cyberbodies, Cyberpunk: Cultures of Technological Embodiment*, ed. Mike Featherstone and Roger Burrows, 65–78. London: Sage, 1995.
Heim, Michael. 'The Feng Shui of Virtual Reality.' *Crossings:e journal of Art and Technology*. Vol 1, June 2001, available from: http://crossings.tcd.ie/issues/1.1/Heim/ [accessed 5 April 2002].
Heilbron, John L. ed. *The Oxford Companion to the History of Modern Science*. Oxford: Oxford University Press, 2003.
Hillis, Daniel. *The Pattern on the Stone*. London: Weidenfield and Nicholson, 1998.
Hillis, Ken. *Digital Sensations – Space, Identity and Embodiment in Virtual Reality*. Minneapolis and London: University of Minnesota Press, 1999.
Holtmeier, Matthew. 'Post-Pandoran Depression or Na'vi Sympathy: Avatar, Affect and Audience Reception.' *Journal for the Study of Religion, Nature and Culture* 4, no. 4 (2010): 414–424.
Holtzman, Steven. *Digital Mosaics: The Aesthetics of Cyberspace*. New York: Simon and Schuster, 1997.
Howarth, David. *Discourse*. Buckingham: Open University Press, 2000.
Hussey, Andrew. *The Game of War: The Life and Death of Guy Debord*. London: Jonathan Cape, 2002.
Huhtamo, Erkki. 'Encapsulated Bodies in Motion: Simulators and the Quest for Total Immersion.' In *Critical Issues in Electronic Media*, ed. Simon Penny, 159–186. New York: University of New York Press, 1995.

Inwood, Michael.J. 'Martin Heidegger.' In *The Oxford Companion to Philosophy*, ed. Ted Honderich. Oxford: Oxford University Press, 2005.

Jameson, Fredric. 'The Cultural Logic of Late Capitalism.' *New Left Review* 1 (1984): 59–92.

Jameson, Fredric. *Postmodernism, or, The Cultural Logic of Late Capitalism*. Durham: Duke University Press, 1991.

Jameson, Fredric. *Archaeologies of the Future*. London: Verso, 2005.

Johnstone, Barbara. *Discourse Analysis*. Oxford: Blackwell, 2002.

Jones, Steve. 'Hyper-Punk: Cyberpunk and Information Technology.' *Journal of Popular Culture* 28, no. 2 (1994): 81–92.

Jones, Thomas. 'ReReading: Beyond Cyberspace.' *The Guardian* Saturday 24th September (2011): 22.

Jordan, Tim. *Cyberpower – The Culture and Politics of Cyberspace and the Internet.* London and New York: Routledge, 1999.

Julier, Guy. 'From Visual Culture to Design Culture.' *Design Issues* 22, no. 1 (2006): 64–76.

Kaplan, Richard. 'Between Mass Society and Revolutionary Praxis: The Contradictions of Guy Debord's Society of the Spectacle.' *European Journal of Cultural Studies* 15, no. 4 (2012): 457–478.

Keen, Andrew. *The Cult of the Amateur*. London: Nicholas Brearley, 2008.

Kellner, Douglas. *Jean Baudrillard: From Marxism to Postmodernism and Beyond.* Oxford: Blackwell, 1989.

Kellner, Douglas. *Baudrillard – A Critical Reader*. Oxford: Blackwell, 1994.

Kellner, Douglas. *Media Spectacle*. London: Routledge, 2002.

Kelly, Kevin. 'Nerd Theology.' *Technology in Society* 21, no. 4 (1999): 387–392.

Kimble, Greg. 'This Is Cinerama.' 2002, available from: http://www.in70mm [accessed 19th January 2012].

Kirkeboen, Geir. 'Descartes' Embodied Psychology: Descartes or Damasio's Error?' *Journal of the History of the Neurosciences* 10, no. 2 (2001): 173–191.

Klein, Herbert. 'From Romanticism to Virtual Reality: Charles Babbage, William Gibson and the Construction of Cyberspace.' *Technoculture, Interdisciplinary Humanities Journal* 24, no. 1 (2007): 36–49.

Klein, Naomi. *No Logo*. London: Flamingo, 2000.

Knowles, Elizabeth. *The Oxford Dictionary of Phrase and Fable*. Oxford: Oxford University Press, 2005.

Kraus, Gerad. 'Video Games: Platforms, Programmes and Players.' In *Digital Cultures – Understanding New Media*, ed. Glenn Creeber and Martin, Royston, 76–85. Maidenhead: Open University Press, 2009.

Kraus, Gerard. 'Video Games: Platforms, Programmes and Players.' In *Digital Cultures – Understanding New Media*, ed. Glen Creeber and Martin Royston, 76–85. Maidenhead: Open University Press, 2009.

Kroker, Arthur. and Michael A. Weinstein. 'DataTrash: The Theory of the Virtual Class.' In *Electronic Media and Technoculture*, ed. John Thornton Caldwell, 117–136. New Brunswick, NJ: Rutgers University Press, 2000.

Kurzweil, Ray. *The Age of Spiritual Machines*. London: Orion, 1999.

Kurzweil, Ray. '*The Human Machine Merger: Are We Headed for The Matrix?' In Taking the Red Pill – Science, Philosophy and Religion in the Matrix*, ed. Glenn Yeffeth, 220–234. Chichester: Summersdale Publishers Ltd, 2003.

Kurzweil, Ray. *The Singularity Is Near*. London: Gerald Duckworth & Co. Ltd, 2009.

Landsberg, Alison. 'Prosthetic Memory: Total Recall and Blade Runner.' In *Liquid Metal – The Science Fiction Reader*, ed. Sean Redmond, 239–248. London and New York: Wallflower, 2004.

Lane, Richard. J. *Jean Baudrillard*. London and New York: Routledge, 2000.

Lanier, Jaron. 'The Future of War: A Farewell to Arms?' *Psychology Today* November–December (1997): 62.

Lanier, Jaron. 'The Third Culture.' *The Edge*. No. 74, 2000, available from: http://www.edge.org [accessed 25 September 2000].

Lanier, Jaron. *You Are Not a Gadget: A Manifesto*. London: Penguin, 2011.

Lauzen, Martha M. 'Kathryn Bigelow: On Her Own in No – (Wo) Man's Land.' *Camera Obscura* 26, no. 78 (2011): 146–153.

Lavik, Erlend. 'The Battle for The Blockbuster: Discourses of Spectacle and Excess.' *New Review of Film and Television* 6, no. 2 (2008): 169–187.

Leary, Timothy. *Chaos & Cyberculture*. Berkley, CA: Ronin, 1994.

Levend, Erlend. 'The Battle for The Blockbuster: Discourses of Spectacle and Excess.' *New Review of Film and Television Studies* 6, no. 2 (2008): 169–187.

Levy, Steven. *Hackers – Heroes of the Computer Revolution*. London: Penguin, 2001.

Lindberg, Kathryne, V. 'Prosthetic Mnemonics and Prophylactic Politics: William Gibson Among The Subjectivity Mechanisms.' *Boundary* 23, no. 2 (1996): 48–83.

Livingstone, E.A. *The Concise Oxford Dictionary of the Christian Church*. Oxford: Oxford University Press, 2006.

Livingstone, Sonia and Leslie Haddon. eds. *Kids Online – Opportunities and Risks for Children*. Cambridge: Polity, 2009.

Livingstone, Sonia, Leslie Haddon, Anke Görzig and Kjartan Ólafsson. (2011) 'Risks and Safety on the Internet: The Perspective of European Children: Full Findings and Policy Implications from the EU Kids Online Survey of 9–16 Year Olds and Their Parents in 25 Countries.' *EU Kids Online*, Deliverable D4. EU Kids Online Network, London, UK.

MacLeod, Douglas and Mary Anne Moser. eds. *Immersed In Technology: Art and Virtual Environments*. Cambridge, MA and London: MIT Press, 1996.

Martellozzo, Elena. *On Line Sexual Abuse: Grooming, Policy and Child Protection in a Multi-Media World*. London: Routledge, 2012.

McCabe, Colin. *Theoretical Essays: Film, Linguistics, Literature*. Manchester: Manchester University Press, 1985.

McCallum, Ellen.L. 'Mapping the Real in Cyberfiction.' *Poetics Today* 21, no. 2, Summer (2000): 349–377.

McCarthy, John and Peter Wright. eds. *Technology as Experience*. Cambridge, MA: MIT Press, 2004.

McCarron, Kevin. 'Corpses, Animals and Mannequins: The Body in Cyberpunk.' In *Cyberspace, Cyberbodies, Cyberpunk*, ed. Mike Featherstone and RogerBurrows, 261–274. London: Sage, 1995.

McRobert, Laurie. *Char Davies Immersive Virtual Art and the Essence of Spatiality*. Toronto: University of Toronto Press, 2007.

Mealing, Stuart. ed. *Computers and Art*. Bristol: Intellect, 2002.

Merrin, William. 'To Play with Phantoms: Jean Baudrillard and the Evil Demon of the Simulacrum.' *Economy and Society* 30, no. 1 (2001): 85–111.

Merritt, Richard K. and Melissa Sopousek. 'Embodiment – The Aesthetics of Incorporation.' *Consumption, Markets and Culture* 5, no. 1 (2002): 47–53.

Miller, Toby. 'Gaming for Beginners.' *Games and Culture* 1, no. 1 (2006): 5–12.

Mirzoeff, Nicholas. *An Introduction to Visual Culture*. London: Routledge, 1999.

Mitchell, William J. *The Reconfigured Eye: Visual Truth in the Post-Photographic Era*. Cambridge, MA: MIT Press, 1992.

Moores, Shaun. *Media, Place and Mobility*. Basingstoke: Palgrave MacMillan, 2012.

Moravec, Hans. *The Future of Robot and Human Intelligence*. Cambridge, MA: Harvard University Press, 1988.

Morley, David and Charlotte Brunsdon. *The Nationwide Television Studies*. London: Routledge, 2005.

Mortensen, Torill Elvira. 'Humans Playing World of Warcraft: Or Deviant Strategies.' In *Digital Culture, Play and Identity*, ed. Hilde Corneliussen and Jill Rettberg Walker, 203–223. Cambridge, MA: MIT Press, 2008.

Mosca, Aldo. 'A Review Essay on Antonio Damasio's The Feeling of What Happens: Body and Emotion in the Making of Consciousness.' *Psyche* 6, no. 10, October (2000) n.p. available from: http://www.theassc.org/vol_6_2000 [accessed 23rd January 2013].

Nakamura, Lisa. *Cybertypes: Race, Ethnicity and Identity on the Internet*. New York: Routledge, 2002.

Nazare, Joseph. 'Marlowe in Mirrorshades: The Cyberpunk (Re) Vision of Chandler.' *Studies in the Novel* 35, no. 3 (2003): 384–404.

Nieborg, David B. 'America's Army: More than a Game?' In *Transforming Knowledge into Action through Gaming and Simulation*, ed. Thomas Eberle and Willy Christian Kriz. Munchen: SAGSAGA, 2004.

Noble, Denis. *The Music of Life: Biology Beyond the Genome*. Oxford: Oxford University Press, 2006.

Norris, Christopher. *Reclaiming Truth – Contribution to a Critique of Cultural Relativism*. London: Lawrence and Wishart Ltd, 1996.

O'Reagan, K.J. and Alva Noe. 'A Sensorimotor Account of Vision.' *Behavioural and Brain: Sciences* 24, no. 5 (2001): 939–1011.

Örtqvist, Daniel and Mats Liljedhal. 'Immersion and Gameplay Experience: A Contingency Framework.' *International Journal of Computer Games Technology* (2010), 1–11.

Papacharissi, Z. 'The Virtual Geographies of Social Networks: A Comparative Analysis of Facebook, LinkedIn and a SmallWorld.' *New Media Society* 1, no. 1–2 (2009): 199–220.

Pariser, Eli. *The Filter Bubble – What the Internet Is Hiding from You*. London: Viking, 2011.

Parkes, Graham. 'Lao-Zhuang and Heidegger on Nature and Technology.' *Journal of Chinese Philosophy* 30, no. 1 (2003): 19–38.

Penny, Simon. *Critical Issues in Electronic Media*. New York: University of New York Press, 1995.

Piaseki Poulsen, Frank. 'Children of the Congo who Risk Their Lives to Supply Our Mobile Phones.' *The Guardian* Friday 7th September (2012).

Plant, Sadie. *Zeros and Ones: Digital Women and the New Technoculture*. London: Fourth Estate, 1997.

Platoni, Karen. 'The Pentagon Goes to the Video Arcade.' *The Progressive* July (1999): 27–30.

Pollock, Griselda. *Vision and Difference: Femininity, Feminism and Histories of Art*. London: Routledge, 1988.

Poole, Steven. *Trigger Happy*. London: Fourth Estate, 2000.

Porter, Stephen. 'Expanding the VR Aesthetic.' *Computer Graphics World* 19, no. 7 (1996): 4.

Poster, Mark. *The Mode of Information: Poststructuralism and Social Context*. Cambridge: Polity, 1990.

Powell, Anna. 'Blood on the Borders: Near Dark and Blue Steel.' *Screen* 35, no. 2 (1994): 136–156.

Prensky, Marc. 'Digital Natives, Digital Immigrants.' *On the Horizon* 9, no. 5 (2001): 1–6.

Purdon, James. 'Quick, Bill-Invent Something!' *The Observer* 12th September (2010): 41.

Ramsay, Morgan. *Gamers at Work*. New York: Springer-Verlag, 2012.

Rascaroli, Laura. 'Steel in the Gaze: On POV and the Discourse of Vision in Kathryn Bigelow's Cinema.' *Screen* 38, no. 3 (1997): 232–246.

Rettberg, Scott. 'Corporate Ideology in World of Warcraft.' In *Digital Culture, Play and Identity*, ed. Hilde Corneliussen and Jill Rettberg Walker, 19–38. Cambridge, MA: MIT Press, 2008.

Rheingold, Howard. *Virtual Reality*. London: Secker and Warburg, 1991.

Rheingold, Howard. *The Virtual Community: Finding Connection in a Computerized World*. London: Secker and Warburg, 1994.

Roberts, John. ed. *Oxford Dictionary of the Classical World*. Oxford: Oxford University Press, 2007.

Robson, Leo. 'Herf Guns in Soho.' *New Statesman* 6th September (2010): 53–54.

Rose, Steven. *The 21st Century Brain*. London: Vintage, 2005.

Rushton, Richard and Gary Bettinson. eds. *What Is Film Theory?*. Maidenhead: Open University, McGraw Hill, 2010.

Rutledge, Virginia. 'Reality by Other Means.' *Art in America* 84, no. 6 June (1996): 38–39.

Schaffer, Simon. 'Babbage's Intelligence: Calculating Engines and the Factory System.' *Critical Inquiry* 21, no. 1 (1994): 203–227.

Sennett, Richard. *The Fall of Public Man*. London: Vintage, 1978.

Shirky, Clay. *Here Comes Everybody – How Change Happens When People Come Together*. London: Penguin, 2009.

Sideris, Lisa H. 'I See You: Interspecies Empathy and Avatar.' *Journal for the Study of Religion, Nature and Culture* 4, no. 4 (2010): 457–477.

Simmons, Dan. '3G Signal Reaches Brazilian Amazon.' *Click* British Broadcasting Corporation, Saturday 22nd September (2012), available from: http://www.bbc.co.uk/click-online/9753821.stm.

Siuru, Bill. 'Virtual Reality for More than Fun and Games.' *Electronics Now* 69, no. 6 (1998): 13–15.

Sobchack, Vivian. 'The Scene of the Screen: Envisioning Cinematic and Electronic "Presence".' In *Materialities of Communication*, ed. Hans Ulrich Gumbrecht and K. Ludwig Pfeiffer, 83–106. Stanford: Stanford University Press, 1994.

Sobchack, Vivian. 'The Scene of the Screen: Envisioning Cinematic and Electronic Presence.' In *Electronic Media and Technoculture*, ed. John Thornton Caldwell, 137–155. New Brunswick, NJ: Rutgers University Press, 2000.

Sobchack, Vivian. 'Postfuturism.' In *Liquid Metal – The Science Fiction Reader*, ed. Sean Redmond, 220–227. London and New York: Wallflower, 2004.

Sohn, Dongyoung. 'Anatomy of Interaction Experience: Distinguishing Sensory, Semantic and Behavioral Dimensions of Interactivity.' *New Media and Society* 13, no. 8 (2011): 1320–1335.

Song, Min Hyoung. *Strange Future – Pessimism and the 1992 Los Angeles Riots*. Durham: Duke University Press, 2005.

Spurling, Laurie. *Phenomenology and the Social World – The Philosophy of Merleau-Ponty and Its Relation to the Social Sciences*. London: Routledge, Kegan Paul, 1977.

Steedman, Carolyn. 'Archival Methods.' In *Research Methods for English Studies*, ed. Gabrielle Griffin, 17–29. Edinburgh: Edinburgh University Press, 2005.

Stephenson, Neal. *Snowcrash*. London: ROC, 1993.

Sterling, Bruce. *Mirrorshades: The Cyberpunk Anthology*. New York: Ace Books, 1986.

Stevens, Tyler. 'Sinister Fruitiness: Neuromancer, Internet Sexuality and the Turing Test.' *Studies in the Novel* 28, no. 3 (1996): 414–433.

Stevenson, Angus. ed. *Oxford English Dictionary*. Third Edition, Oxford: Oxford University Press, 2010.

Stone, Allucquére Rosanne. *The War of Desire and Technology at the Close of the Mechanical Age*. Cambridge, MA: MIT Press, 1995.

Strayer, Kirsten. 'Reinventing the Inhuman: Avatars, Cylons, and Homo Sapiens in Contemporary Science-Fiction Television Series.' *Literature Film Quarterly* 38, no. 3 (2010): 194–204.

Sturken, Marita and Lisa Cartwright. *Practices of Looking :An Introduction to Visual Culture*. Oxford: Oxford University Press, 2001.

Sutherland, Ivan. *Sketchpad: A Man-Machine Graphical Communication System*. Cambridge, MA: MIT Doctoral Thesis, 1963.

Talbott, Stephen.L. *The Future Does Not Compute: Transcending the Machines in Our Midst*. California, CA: O'Reilly Associates Inc, 1995.

Tasker, Yvonne. *Spectacular Bodies, Gender, Genre and the Action Cinema*. London and New York: Routledge, 1993.

Tenhaaf, Nell. 'As Art Is Lifelike – Evolution, Art and the Readymade.' *Leonardo* 31, no. 5 (1999): 397–404.

Thrift, Nigel. 'Summoning Life.' In *Visioning Human Geographies*, ed. Paul Cloke, Philip Crang and Mark Goodwin, 81–30. Arnold: London, 2002.

Thrift, Nigel. 'Movement-Space: The Changing Domain of Thinking Resulting From the Development of New Kinds of Spatial Awareness.' *Economy and Society* 33, no. 4 (2004): 582–604.

Thrift, Nigel. 'From Born to Made: Technology, Biology and Space.' *Transactions of the British Institute of Geographers* 30 (2005): 463–476.

Thrift, Nigel. *Non-Representational Theory: Space, Politics, Affect*. London: Routledge, 2008.

Toles, George. 'Rescuing Fragments: A New Task for Cinephilia.' *Cinema Journal* 49, Winter (2010): 159–166.

Tran, Mark 'Girl Starved to Death While Parents Raised Virtual Child in Online Game.' *The Guardian*, 5th March 2010, available from: http://www.guardian.co.uk [accessed 30th August 2011].

Turkle, Sherry. 'Constructions and Reconstructions of Self in Virtual Reality: Playing in the MUDS.' *Mind, Culture and Activity* 1, no. 3 (1994): 158–167.

Turkle, Sherry. *Life on Screen: Identity in the Age of the Internet*. New York: Simon and Schuster, 1997.

Turkle, Sherry. *Alone Together*. New York: Basic Books, 2011.

Turner, Fred. 'How Digital Technology Found Utopian Ideology: Lessons from the First Hackers Conference.' In *Critical Cyber-Culture Studies*, ed. David Silver and Adrienne. Massanari, 257–269. New York: New York Uni Press, 2006.

Van Gelder, Lindsy. 'The Strange Case of the Electronic Lover.' In *Computerization and Controversy*, ed. Chuck Dunlop and Rob Kling, 533–546. San Diego: Academic Press, 1991.

Van Gelder, Lindsy. 'The Strange Case of the Electronic Lover.' In *Computerization and Controversy*, ed. Rob Kling, 533–546. Orlando: Academic Press, 1996.

Virilio, Paul. *Open Sky*. Translated by Julie Rose. London and New York: Verso, 2000.

Wade, Robert H. and Silla Sigurgeirsdottir. 'Iceland's Meltdown: The Rise and Fall of International Banking in the North Atlantic.' *Brazilian Journal of Political Economy* 31, no. 5 (2011): 684–698.

Wallace-Wells, David. 'William Gibson, The Art of Fiction.' *The Paris Review* 197, Summer 2011, available from: http://www.theparisreview.org [accessed 19 June 2012].

Wallace-Wells, David. 'William Gibson, The Art of Fiction No. 211.' *The Paris Review*, Summer, 2011, available from http://www.theparisreview.org/interviews/6089/the-art-of-fiction-no-211-william-gibson [accessed 20th August 2013]

Waller, Fred. 'The Archeology of Cinerama.' *Film History* 5, no. 3 (1993): 289–297.

Weatherall, James Owen. *The Physics of Wall Street*. Boston: Houghton Mifflin Harcourt, 2013.

Wertheim, Margaret. *The Pearly Gates of Cyberspace – A History of Space from Dante to the Internet*. London: Virago, 1999.

Whetmore, Alex. 'The Poetics of Pattern Recognition: William Gibson's Shifting Technological Subject' *Bulletin of Science, Technology and Society* 27. no. 1 (2007): 71–80.

White, Hayden. *Tropics of Discourse-Essays in Cultural Criticism*. Baltimore and London: The John Hopkins University Press, 1978.

Wiener, Norbert. *The Human Use of Human Beings*. Sphere Books: London, 1968.

Wilson, Edward O. *In Search of Nature*. Washington: Island Press, 1997.

Wilson, Edward O. *The Diversity of Life*. London: Penguin, 2001.

Wooldbridge, Dean. *The Machinery of the Brain*. New York: McGraw Hill, 1968.

Woolley, Benjamin. *Virtual Worlds*. London: Penguin, 1993.

Yasuo, Yuasa. 'Image-Thinking and the Understanding of Being: The Psychological Basis of Linguistic Expression.' *Philosophy East and West* 55, no. 2 (2005): 179–208.

Zurbrugg, Nicholas. ed. *Jean Baudrillard – Art and Artefact*. London: Sage, 1999.

Filmography

Blade Runner (1982) directed by Ridley Scott.

Blood in the Mobile (2010) directed by Frank Piaseki Poulson.

Blue Steel (1989) directed by Kathryn Bigelow.

Brainstorm (1983) directed by Douglas Trumbull.

Catfish (2010) directed by Henry Joost and Ariel Schulman.

Dark City (1998) directed by Alex Proyas.

eXistenz (1999) directed by David Cronenberg.

Ghost in the Shell (1995) directed by Mamoru Oshii.

Gladiator (2000) directed by Ridley Scott.

Groundhog Day (1993) directed by Harold Ramis.

Independence Day (1996) directed by Roland Emmerich.

Jurassic Park (1993) directed by Steven Spielberg.

Life of Pi (2013) directed by Ang Lee.

Prometheus (2012) directed by Ridley Scott.

Sky Captain and the World of Tomorrow (2004) directed by Kerry Conran.

Source Code (2011) directed by Duncan Jones.

Strange Days (1995) directed by Kathryn Bigelow.

Terminator 2 (1991) directed by James Cameron.

The Avengers (2012) directed by Joss Whedon.

The Green Hornet (2011) directed by Michel Gondry.

The Hurt Locker (2008) directed by Kathryn Bigelow.

The Lawnmower Man (1992) directed by Brett Leonard.

The Matrix (1999) directed by Andy and Larry Wachowski.

The Terminator (1984) directed by James Cameron.

The Thirteenth Floor (1999) directed by Josef Rusnak.

The Truman Show (1998) directed by Peter Weir.

Total Recall (1990) directed by Paul Verhoeven.

Tron (1982) directed by Steven Lisberger.

Videodrome (1983) directed by David Cronenberg.

Virtuosity (1995) directed by Brett Leonard.

War Games (1982) directed by John Badham.

Index

www.ingramcontent.com/pod-product-compliance
Lightning Source LLC
Chambersburg PA
CBHW031415220926
51612CB00004B/42
* 9 7 8 1 4 4 1 1 7 5 3 1 1 *